Build Your Own
Quadcopter

Build Your Own Quadcopter

Power Up Your Designs with the Parallax Elev-8

Donald Norris

New York   Chicago   San Francisco
Athens   London   Madrid
Mexico City   Milan   New Delhi
Singapore   Sydney   Toronto

McGraw-Hill Education books are available at special quantity discounts to use as premiums and sales promotions or for use in corporate training programs. To contact a representative, please visit the Contact Us page at www.mhprofessional.com.

**Build Your Own Quadcopter:
Power Up Your Designs with the Parallax Elev-8**

1 2 3 4 5 6 7 8 9 0 DOC/DOC 1 2 0 9 8 7 6 5 4

ISBN   978-0-07-182228-2
MHID   0-07-182228-3

This book is printed on acid-free paper.

Sponsoring Editor
Roger Stewart

Editing Supervisor
Stephen M. Smith

Production Supervisor
Pamela A. Pelton

Acquisitions Coordinator
Amy Stonebraker

Project Manager
Nancy Dimitry,
D&P Editorial Services

Copy Editor
Joseph Cavanagh,
D&P Editorial Services

Proofreader
Don Dimitry,
D&P Editorial Services

Indexer
WordCo Indexing Services

Art Director, Cover
Jeff Weeks

Composition
D&P Editorial Services

To my wife Karen for her continuing and unwavering love and support despite the onslaught of spontaneous lectures from yours truly.

To my children Shauna, Heath, and Derek who, while loving and supporting me, would never put up with my lectures as adults.

About the Author

Donald Norris has a degree in electrical engineering and an MBA specializing in production management. He is currently teaching undergrad and grad courses in the IT subject area at Southern New Hampshire University. He has also created and taught several robotics courses there. He has over 30 years of teaching experience as an adjunct professor at a variety of colleges and universities.

Mr. Norris retired from civilian government service with the U.S. Navy, where he specialized in acoustics related to nuclear submarines and associated advanced digital signal processing. Since then, he has spent more than 17 years as a professional software developer using C, C#, C++, Python, and Java, as well as 5 years as a certified IT security consultant.

Mr. Norris started a consultancy, Norris Embedded Software Solutions (dba NESS LLC), which specializes in developing application solutions using microprocessors and microcontrollers. He likes to think of himself as a perpetual hobbyist and geek, and is always trying out new approaches and out-of-the-box experiments. He is a licensed private pilot, active member of the Civil Air Patrol, photography buff, amateur radio operator, and avid runner.

Mr. Norris is also the author of the TAB McGraw-Hill book *Raspberry Pi Projects for the Evil Genius.*

Contents

Preface

I t was a bit of a surprise to me when my editor, Roger Stewart, asked me if I was interested in writing a book about building and flying quadcopters. It seems Roger had been interested in having a book written about this hot topic for some time. I remember mentioning to him that I had recently built a good-sized quadcopter, just because it was a fun thing to do, and I was quite interested in the technology that permitted these aircraft to not only fly but also be manually controlled precisely, or even be set to fly autonomously. I accepted his offer, and the book you now have in your hands or are viewing on an electronic device is the result of that chance discussion.

I must admit I am somewhat of a multirotor geek hobbyist, having two Parallax Elev-8 units, one octocopter and one micro-sized quadcopter. I must also warn you that this hobby is addictive and you will soon see yourself surrounded by quadcopters and, more likely, pieces and parts of quadcopters, which is just part of the price you pay to enjoy this hobby. But have no fear, it is perfectly possible to minimize damage yet thoroughly enjoy flying your quadcopter for many, many enjoyable hours.

I would like now to give you my high-level view (pun intended) of this book and what I hope you will gain from reading it. First, and most likely foremost, in the minds of readers is that you will be able to successfully build the Parallax Corporation's Elev-8 quadcopter kit by following the instructions in this book, which are detailed in Chapter 3. I will honestly say that these instructions are mainly from the Parallax assembly instructions with plenty of additional information that I have provided to clarify and amplify the company-provided instructions. Having said that, I will emphasis that this book is a lot more than a "missing manual"–type book. As an educator, I feel somewhat responsible that my readers get not only what is needed to build a quadcopter but also a lot more in terms of an education about all the technologies that make up today's sophisticated quadcopters. With that in mind, I will explain all the principal components that constitute a modern quadcopter in sufficient detail so that you should feel comfortable deciding which components you can or should modify to suit your own needs and desires. I would also like to mention that the quadcopter that is being discussed in this book would likely be classified as professional or semiprofessional in nature so as to be distinguished from the flood of very cheap manufactured "toy" quadcopters. I am not being derisive toward the toys, as they have their place, but it is totally wrong to lump the two types together, as you will see as you progress through this book. Now, I will gently step down from my "soapbox" and proceed to tell you what to expect in this book.

The first chapter starts with a history of the quadcopter or, as it was known in historical times, a multirotor aircraft. Back in the 1920s, there was no concept of unmanned aircraft; therefore, all the experimental multirotors were large-scale units, fully capable of carrying one person airborne. How those flights turned out are other stories you will find in the chapter. I then progress rapidly through the twentieth century to the 1990s, where rapid progress, mainly in terms of semiconductor and battery technologies, makes the modern quadcopter a real possibility.

Truly understanding how a quadcopter flies was my reason for including Chapter 2, which covers basic flight aerodynamics. Make no mistake; quadcopters are governed by the same flight principles that apply from the Wright brothers' "Flyer" to the ultramodern F-35 *Joint Strike Fighter* (JSF). As a real pilot, I thought it was important that readers understand how the quadcopter can be made to fly and what aerodynamic forces are constantly in play while it is flying. Along the way, I threw in a "little" math in terms of *proportional-integral-derivative* (PID) theory to provide a basis for understanding the control protocol or algorithms that are needed to keep the quadcopter in a steady flight pattern.

Chapter 3 contains all the build instructions for assembling and configuring the Parallax Corporation's Elev-8 quadcopter kit. Go no further than this chapter if you simply want to read an expanded set of assembly instructions. However, I expect my readers to be far more interested in what makes up the quadcopter system and will read on. I promise you that you will not be disappointed.

The Propeller chip supplied by Parallax is the heart of the quadcopter flight-control board. Chapter 4 explores what constitutes this fantastic technology and how you can learn to program this chip to perform the experimental functions you can invent and desire to test. I also introduce and explain the concept of *pulse-width modulation* (PWM), which is an integral technology needed to control the quadcopter.

Chapter 5 covers all the propulsor components that make up larger-scale quadcopters, including the motors, electronic speed controllers (ESCs) and the propellers. All of these are essential parts of the quadcopter's propulsion system, and it is very important that you understand what limitations and constraints apply to each of them. "Overtaxing" your motors will cause the quadcopter to fail, and probably at the worst time possible.

The next chapter covers radio-controlled (R/C) systems. Don't worry, you will not be required to build your own; however, I do want you to understand why certain R/C systems are so much better than others that are normally much less expensive. I make the case that it is a wise investment to acquire a high-quality R/C system to ensure that you maintain positive control over your quadcopter at all times. There are many inexpensive systems available, and while they may be satisfactory for toy systems, they really are not suitable for a relatively expensive and larger-scale quadcopter like the Elev-8. I also show you how to program a Parallax development board to measure certain key signals that transmit from your R/C system.

Chapter 7 covers R/C grade servos, which is a bit odd, as the basic Elev-8 kit does not contain any servos. I included this servo chapter to ensure that you are well acquainted with this technology, as these devices are used extensively in "regular" R/C aircraft and also in a modification to the Elev-8 for controlling the tilt of an onboard video camera. I also show how to build an LED-flasher circuit that takes advantage of a spare servo channel that is available on the quality R/C transmitter.

GPS is covered in Chapter 8, in which I start with what I hope is a good, but brief, tutorial on how GPS functions and how it could be used in conjunction with quadcopter operations. I also show you how to build a real-time GPS data-reporting system, using XBee technology to transmit data from the quadcopter to a *ground control station* (GCS). Theoretically, you could control the quadcopter using the transmitted GPS coordinates well beyond the R/C operator's *line-of-sight* (LOS); however, I strongly do not recommend this mode of operation.

I discuss airborne video systems in Chapter 9, as that is truly a hot-topic item with regard to quadcopters. Two types of video systems are shown, one which provides high-quality, wide-angle views, and another which is much lower quality but still more than adequate to be used with video-processing software, which is also discussed in the chapter. Although not mentioned in the chapter, I do want to acknowledge that I have been involved with an experimental quadcopter video surveillance system deployed on or about my college campus. I also want to acknowledge the help and support of Dr. Lundy Lewis in this project, which was designed to promote a campus-wide, reasoned discussion of both the advantages and disadvantages of deploying such a system.

Training is the chief topic in Chapter 10. Learning how to safely control a quadcopter is a necessity; it definitely requires patience and repeated use of some type of simulator before you acquire the skills to safely fly the Elev-8. This is another area that differentiates the toy class from the professional-grade quadcopter. Most people can learn to control the toy versions by trial and error without incurring much damage to the toy or endangering other people or property. That's not true with the Elev-8; you should practice and hone your skills before attempting to fly it, especially in congested areas.

The final chapter deals with further enhancements to the Elev-8 and suggestions for future projects that might interest readers, such as autonomous flight and applying *artificial intelligence* (AI) to quadcopter operations. I discuss an electronic compass sensor, which is an important add-on, especially if autonomous operations are being considered. A brief introduction to *Fuzzy Logic* (FL) is also presented in the chapter, as it is the most appropriate control approach needed to implement quadcopter AI. I would also like to acknowledge the support of Dr. Robert Seidman, who taught me so much about AI and how to properly apply it to control scenarios, which is so applicable in this situation.

I hope the book only opens your desire to participate in building and flying a quadcopter. Quadcoptera are much more than a simple hobby, as you probably realize from reading current articles and seeing TV news segments. The overall "drone" market is expected to grow into the multibillion range in the very near future, and, hopefully, this book will help you prepare to participate in this rapidly changing, but highly interesting, endeavor/hobby.

Good luck with your quadcopter.

Donald Norris

Build Your Own Quadcopter

CHAPTER 1

Introduction to Quadcopters

A Brief History of Multirotor Helicopters

The multirotor helicopter also known as a *quadrotor* or *quadcopter* is equipped with four rotors to create lift. It is a true helicopter in that lift force is created by narrow-chord horizontally rotating air foils. The quadcopter design has been in existence since the 1920s when an early manned version named the De Bothezat helicopter was built and successfully flown. First developed and prototyped under a U.S. Army contract, the De Bothezat helicopter is pictured in Figure 1.1.

It first flew in October 1922 at what is now known as Wright Field in Dayton, Ohio. The helicopter actually started with six rotors, but eventually two were deemed unnecessary and were eliminated. It made more than 100 flights over a period of years but never flew more than 5 meters into the air and never with any lateral movement. This was due to the complexity and difficulty of simply trying to maintain level flight, never mind moving in a lateral direction. This lateral movement control was to be the bane of multirotor helicopters until the invention and use of computer-assisted flight-control systems that would lessen the pilot workload. The U.S. Army eventually lost interest in the De Bothezat project and discontinued it in the early 1930s, after spending more than $200,000 on the program.

Helicopter development languished, at least in the United States, from the early 1930s to the mid 1940s. With the ending of World War II, development work did resume, but the focus was on more conventional designs that employed a main rotor with a tail rotor or the use of coaxial main rotors. The armed forces that initially funded helicopter development apparently believed that any possible advantages of using quad rotors were far outweighed by their complexity and ill-mannered flight characteristics.

The U.S. Army eventually developed and successfully fielded a heavy-lift, tandem-rotor helicopter named the Chinook, model CH-47, which despite being designed in the 1960s, is still in wide use today. It has undergone many updates and upgrades to keep it fully compatible with today's environment.

The U.S. Department of Defense also sponsored the development and production of a hybrid, dual-tilt-rotor aircraft named the Osprey, model V-22. It takes off and lands as a dual rotor helicopter, but flies as a traditional airplane with the wings tilted to a level position while it is operating in cruise mode. Figure 1.2 is a picture of the pilot's station in the V-22, showing all the incredible technology available to the pilot.

FIGURE 1.1   De Bothezat helicopter.

Both the Chinook and Osprey take advantage of computer-assisted flight-control systems that significantly reduce pilot workload and make it practical to safely fly aircraft that would otherwise be nearly impossible to fly.

The development of true quad-rotor helicopters turned out to be delayed until the early 1990s when a small-scale, *radio-controlled* (R/C) system named the Gyro Saucer 1 was developed and marketed in Japan. This is the earliest instance that I could find in my research for the appearance of a practical quadcopter, with or without an onboard pilot. It used mechanical gyros for stability and fairly small electrical motors to turn the props. Unfortunately, the props were made of Styrofoam and had a habit of disintegrating if they came in contact with anything, including light fabric curtains. The Gyro Saucer had an operating flight time of approximately three minutes, was never exported from Japan, and hence, was a relatively unknown system. Figure 1.3 is a picture of this early quadcopter.

The first modern, widely available multirotor system was the Draganflyer, which was designed and manufactured in the early 2000s by Draganfly Innovations Inc. Draganfly has

FIGURE 1.2   V-22 pilot's station.

Figure 1.3   Gyro Saucer 1 system.

since superseded that early design with later models that are much more sophisticated and come equipped with a variety of functional capabilities. Figure 1.4 is a picture of their X-8 model, which is quite a remarkable and stable platform.

The X-8 quadcopter has four booms with a motor attached to each one and a pair of propellers attached to each motor, thus making for a total of eight propellers on the craft. This quadcopter is just one of dozens of models available for purchase at the time this book is being written.

Most small-scale, R/C multirotor helicopters have four rotors; however, there are models with as few as three to as many as eight, with a few outliers with even more. There is also a start-up company named *e-volo* that plans to build a manned aircraft with 18 rotors named the *Volocopter*.

This book will focus only on building and flying an R/C small-scale quadcopter because it is the most representative and reasonably priced of the current selection of multirotor helicopters.

Figure 1.4   Draganflyer X-8.

A Matter of Definition

There are a variety of descriptors associated with quadcopters that I would like to briefly examine. Probably the most general description of a quadcopter is an *unmanned aerial vehicle* (UAV). UAV has also been interpreted as an *uninhabited aerial vehicle*, which is precisely the same meaning as unmanned aerial vehicle. Two descriptions that are more specific would be those of *remotely operated aircraft* (ROA) and *remotely piloted vehicle* (RPV). The last two descriptions mean that no pilot is physically carried by the aerial vehicle and all vehicle control is accomplished either by a pilot using a remote ground station or autonomously by the vehicle. A related definition for this type of operation is *autonomous aerial vehicle* (AAV). It is often used to describe a UAV that is controlling its own operation, independent of any ground station. However, it should be pointed out that all AAVs should have some kind of autonomous override command available from a ground station, just in case something goes wrong with the onboard flight-control system. Having a fail-safe mode should always be a paramount design decision in any AAV project.

UAV, ROA, and RPV are the most popular and well-known descriptions for the quadcopter type of vehicle. There is also one other popular descriptor: *micro aerial vehicle* (MAV), which refers to any very small-sized UAV with all length, width, or height dimensions of 15 cm or less. MAVs are actively being developed along with swarming control techniques in a variety of research projects. Developers of some of these projects hope to implement insect biomimicry into their MAVs in an attempt to achieve the performance and capabilities of their real-world counterparts.

How Are Quadcopters Used?

The answer to the question of how quadcopters are used depends on whether one is viewing them from a military or a civilian perspective. Military use of quadcopters lies mainly in the intelligence, surveillance, and reconnaissance (ISR) field, and to a lesser extent, in tactical deployments. Quadcopters are excellent ISR assets that nicely complement fixed-wing UAVs that are extensively employed by many worldwide military organizations. Currently all tactical deployments where weapons-carrying UAVs are deployed in actual combat are still the domain of fixed wing UAVs, such as the U.S. Air Force MQ-9 Reaper, which is shown in Figure 1.5.

The quadcopter, as of this writing, still cannot carry a heavy payload, such as a missile or cannon, although it is likely that there are ongoing military research projects attempting to overcome this limitation. There would be an obvious tactical advantage for a combat unit to be able to deploy a small, airborne weapons platform that could hover over a battlefield and engage enemy targets upon command. The old military adage of "gain the high ground for tactical advantage" would definitely take on a new meaning with the use of a tactical quadcopter.

Civilian uses for quadcopters are far more numerous than military ones at this time. Some of these are listed in Table 1.1.

There are some legal restrictions in the United States regarding the civilian use of quadcopter UAVs, including the FAA requirements that they not be flown any more than 400 feet above ground level and not near any airport. I am sure that similar restrictions are in place in other countries, so I would urge you to research the laws and regulations that are applicable in your country.

U.S. residents should also be aware that the airspace above their domiciles is *not* exclusive for their use. In the 1946 decision, United States v. Causby 328 US 256 1946, the U.S. Supreme Court held that only the landowner's airspace that may be reasonably occupied or

FIGURE 1.5 US Air Force MQ-9 Reaper UAV.

used in connection with the land residence is exclusive to the landowner. The precise court wording is shown below:

> Cujus est solum ejus est usque ad coelum et ad inferos *has no legal authority in the United States when pertaining to the sky. A man does not have control and ownership over the airspace of their property except within reasonable limits to utilize their property. Airspace above a set minimum height is property of the Masses and no one man can accuse airplanes or other such craft of trespassing on what they own.*

Law enforcement
Security patrols on private property
Agricultural surveying
Communications relay
Incident command support
Aerial mapping
Aerial photography
Severe weather telemetry
University research projects
Search and rescue

TABLE 1.1 Civilian Uses of Quadcopters

The Latin words at the start of the court's decision refer to English common law where it was held that a landowner had exclusive rights to all space "from the depths to the heavens." Obviously, the court held that this specificity did not apply in the United States. Otherwise, one could imagine the resulting chaos if airlines had to obtain landowners permission to fly into airspace that projected from the ground. As mentioned earlier, it would be wise to check with the appropriate authorities before flying your quadcopter in a country other than the United States. You might find yourself inadvertently trespassing in someone's airspace.

Other limitations or constraints related to real-time video surveillance are more problematic. All quadcopters with video capability, whether onboard capture or real-time transmission, should be operated with prudence. In other words, it is definitely *not* a good idea to fly the quadcopter (even without video) over to your neighbor's house and attempt to peer in their windows. Flying over your neighbors' houses, while legal, should be done with an abundance of caution. I would definitely talk with my neighbors prior to making any flights above or close to their homes.

Design of the Elev-8 Quadcopter

According to Ken Gracey, President of Parallax Inc., the Elev-8 project began after a visit to his company by some folks from the Hoverfly Company. Hoverfly manufactures sophisticated quad and hex copters that can optionally be equipped with camera systems. Hoverfly also designs and manufactures flight-control boards, which was one of the main reasons they visited Parallax. It turns out that some very bright Parallax engineers designed an unusual and very clever eight-core microcontroller they aptly named *Propeller*. The designers decided to call their cores "cogs," which I suppose was to emphasize a more collaborative computing approach as compared to traditional multicore processors. (In a later chapter, I will explore the Propeller chip in much greater depth.) Designers and engineers at Hoverfly recognized the unique capabilities of the Propeller chip and decided to incorporate it into their flight-controller boards. Thus, the reason for the visit to Parallax headquarters was to demonstrate their quadcopter. Ken was fascinated with their demonstration and quickly realized that he and his company had to be involved in a like-minded project, which is the genesis of the Elev-8. Ken also realized that it made much more sense to provide a kit of parts in lieu of a fully assembled quadcopter. This idea fit with the Parallax company specialty, which is centered on providing builders and users with components and subassemblies in lieu of fully assembled products. At times, they have provided fully assembled products, but that seems mostly outside of their modus operandi.

Creating the basic Elev-8 kit was in itself a bit of a problem: Ken and two of his engineers, Kevin Cook and Nick Ernst, had to determine suitable components that would enable builders to successfully make their own Elev-8 without excessive costs or complexity. Many of the problems and design decisions they encountered will be discussed in later chapters to provide you with an understanding of the decisions that are required in a project of this complexity.

It was an easy decision for Ken to simply incorporate a fully assembled HoverflySPORT controller board into the kit. The flight-controller board is the key element that enables a user with an R/C transmitter to fly a quadcopter as directed. Figure 1.6 shows the HoverflySPORT controller board.

Chapter 2 delves into the complexities of quadcopter flight dynamics, and it soon will become apparent that designing and building a flight-control board is best left to professionals.

FIGURE 1.6   HoverflySPORT controller board.

Having said that, it turns out that current Elev-8 kits now contain the HoverflyOPEN controller board, which gives knowledgeable users the opportunity to add their own control programs in lieu of using the default software. Figure 1.7 shows the HoverflyOPEN board In Chapter 2, I will also address the pros and cons of creating your own flight-control software.

Main Electrical/Electronic Elev-8 Components

The main electrical/electronic components that make up the Elev-8 system are shown in Figure 1.8. There are only 11 essential components, not counting wires, connectors, or any optional components such as telemetry and LED display components.

With only 11 components, the Elev-8 is not a very complex aircraft due mainly to the automated control provided by the HoverflyOPEN control board. A fully assembled, basic Elev-8 is shown in Figure 1.9.

The HoverflyOPEN control board and a Spektrum AR8000 receiver are clearly visible mounted on the top of the quadcopter. The LiPo battery is disconnected, as shown by the unconnected power cables that are visible in the front of the quadcopter. Red checkerboard pattern decals have been applied on each of the two aluminum tubes attached to the right side of the quadcopter. In addition, black checkerboard pattern decals are on the tubes

Figure 1.7　HoverflyOPEN controller board.

attached on the left. The red decals serve a very important purpose: they show the forward travel direction for an X configuration quadcopter. Forward is always between the red checkered tubes. The X configuration as well as other configurations will be discussed in Chapter 2.

Figure 1.10 is a picture of the first Elev-8 that I built in early 2012.

Every Elev-8 built will be unique to some degree. They all start from a basic kit of parts available for purchase from Parallax Inc., Rocklin, CA. Users can, and probably should, modify their kits to suit their personal preferences. Modifications can include adding items such as LED lights, video cameras, GPS trackers, and so on. For example, I added a separate Basic Stamp II microcontroller on my first quadcopter, which allowed me to independently program the display operation of the four LED strips attached to the underside of each of the four boom tubes. Figure 1.11 is a picture of a Basic Stamp II development board mounted

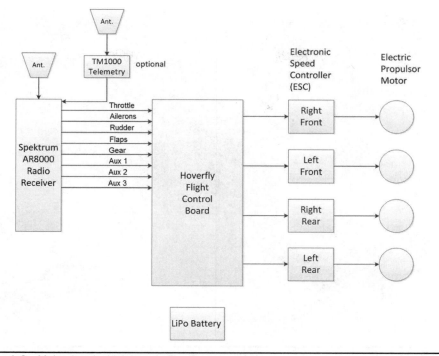

FIGURE 1.8   Main electrical/electronic Elev-8 components.

FIGURE 1.9   Basic Elev-8 quadcopter.

FIGURE 1.10 My first Elev-8 quadcopter.

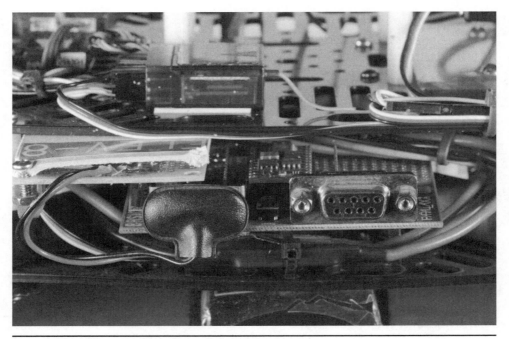

FIGURE 1.11 LED control Basic Stamp II development board.

FIGURE 1.12   LED strip.

between the two Delrin boards that comprise the main structural boards of the Elev-8. The LED power distribution prototype board is located beneath a small cardboard piece with Elev-8 printed on it.

The DB-9 connector visible in the photo is used only to program the Stamp chip and is not needed during normal operation.

One of the LED strips mounted on the bottom of a boom tube is shown in Figure 1.12. Each strip has six LEDs mounted on an integral plastic backing strip that needs only two wires to power and control it.

I also added an R/C servo-control kill switch that was the ultimate fail-safe feature, just in case the quadcopter went out of control. This servo-control switch is shown in Figure 1.13.

Remotely activating this switch immediately cuts off all power to the quadcopter and causes it to drop to the ground. Remember, it is always preferable, and a lot less expensive, to accept damage to the quadcopter than to cause personal injury and/or property damage to an innocent party.

FIGURE 1.13   Servo-control kill switch.

Summary

I began this chapter with a brief history of multirotor development that started in the 1920s with a large manned aircraft, the De Bothezat helicopter. It turned out that safely controlling the flight path for this type of aircraft was much too demanding for human pilots. This fact delayed development until computer technologies were created that enabled the requisite supplemental control for safe flight. Two significant paths of development then led to more advances: (1) amply funded military projects in the 1970s yielded the design and production of manned multirotor aircraft, and (2) small unmanned R/C multirotors were built in Japan in the 1990s.

Next, I discussed a variety of definitions to help clarify the confusion that seems to surround these aircraft. The term unmanned aerial vehicle (UAV) appears to be the most appropriate descriptor for the quadcopter.

The section on military and civilian uses of quadcopters revealed that there are far more civilian applications than military ones. I also discussed some legalities that you should be aware of and obey as required.

The Elev-8 quadcopter kit origins began when the Hoverfly Company used the Parallax Propeller chip in the design of their flight controllers. The president of Parallax decided to create a parts kit, including a Hoverfly control board that would enable users to build their own highly capable quadcopter at a reasonable cost. Remarkably, there are only 11 main electrical/electronic components that make up the basic Elev-8.

I finished the chapter by showing some of the add-ons and enhancements that can be incorporated into the basic Elev-8. A real-time video add-on will also be discussed in detail in a later chapter.

Chapter 2 provides a thorough discussion of the quadcopter flight dynamics. I strongly urge you to carefully study the next chapter in order to achieve a good understanding of the physics that allows a quadcopter to fly. This knowledge will improve your control skills. Additionally, a good understanding of the basics of quadcopter control will also help you to create your own software, if you are so inclined.

CHAPTER 2

Quadcopter Flight Dynamics

Flight Basics

I will begin this chapter with an introduction to basic flight principles that are applicable to any aircraft with wings. It is important for you to have this knowledge in order to understand how a quadcopter flies and the differences between its flight characteristics and those of a normal aircraft.

Figure 2.1, which is from NASA, shows an iconic Wright 1903 Flyer with all four aerodynamic flight forces that simultaneously and continually act upon it. The four forces shown in the figure are further described in Table 2.1.

These forces are universally applicable to all aircraft—from the Wright Flyer to the modern F-35 Joint Strike Fighter (JSF). How an aircraft responds to these forces determines whether it is climbing, diving, in level flight, or turning.

The quadcopter design is different from that of regular aircraft in that it has no wings, and thus, cannot generate any lift force. Instead, it depends solely on thrust forces created by the motors attached at the end of each of its booms. Additionally, the upward and forward velocities traveled by a quadcopter are small enough that drag forces are not really a factor. As a consequence, there are only two principal forces affecting the quadcopter: thrust and weight. Now weight is a fixed force that can be changed only by design or by altering the payload. This leaves only thrust as the sole control force for a quadcopter. However, thrust is nearly directly proportional to the rotational speed of the motors, which means that controlling the motor speed totally controls the flight path of the quadcopter. When the rotational speeds are all equal and sufficiently fast, then the quadcopter will rise straight up into the air. A vertical flight path was essentially the only flight path available to the early De Bothezat helicopter discussed in Chapter 1. Varying the rotational speeds of one or more of the quadcopter motors is the only way to alter the quadcopter's flight path. Altering the quadcopter flight path would be a most daunting proposition for a human pilot who would have to rely on his or her sense of balance and then somehow translate that sensation to actual motor speed changes. It is easy to understand why manned multirotors remained an unachievable goal until the advent of automated flight-control techniques.

Also important to quadcopter flight dynamics are "two additional flight principles of balance and center of gravity (CG), which are directly related to weight, one of the fundamental flight forces. Weight must be properly distributed in order for any aircraft to fly safely. Determining safe weight distribution starts with the basic aircraft design and uses a

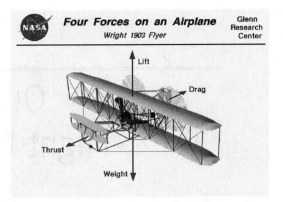

Figure 2.1   Wright 1903 Flyer shown with flight forces. *(Courtesy of NASA)*

center-of-gravity concept. CG can be thought of as an imaginary point within an aircraft where it could be suspended in a perfectly balanced position. In the real world, a CG point is used to determine if an aircraft is stable; if the entire payload including the airframe, fuel, passengers, and cargo is within prescribed design limits; and consequently, if the aircraft is safe to fly. A quadcopter CG may be thought of as the point within the copter where a string could be attached to suspend it in a perfectly balanced state. Naturally, one would expect the CG to be collocated with the physical center of the quadcopter. If the CG is located off center, it would tend to make the quadcopter unstable, perhaps to the point of being uncontrollable if it were located too far from the physical center. This is always something to consider when attaching devices to the quadcopter. For instance, attaching a camera module close to the outboard side of a motor boom, which might make sense for a better field of view, would probably upset the CG to the detriment of the quadcopter flight stability.

Flight Axes

In order to fully understand airplane flight dynamics, it is necessary to discuss three physical axes and the three rotations associated with those axes. Figure 2.2 shows a light *general-aviation* (GA) airplane with its longitudinal axis running fore and aft through the fuselage. The lateral axis is perpendicular and in the same plane as the longitudinal axis and runs through the wing, intersecting with the longitudinal axis at the CG. The third axis, called the vertical axis, is perpendicular to the other two and also goes through the CG. The three

Force	Description
Weight	The downward force acting upon the aircraft due to Earth's gravity.
Lift	The upward force created by the rapid passage of air over and under the airfoil (wing).
Thrust	The forward force created by the rotating propeller pushing air backward.
Drag	The backward force created by wind resistance due to the fuselage shape and non-streamlined appendages.

Table 2.1   Four Aerodynamic Flight Forces

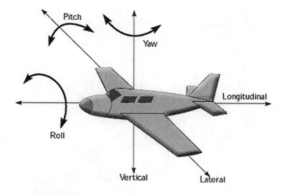

FIGURE 2.2   Airplane principal axes and axial rotation motions.

rotational motions associated with each of these axes are also shown in Figure 2.2 and are described in Table 2.2.

Actual flight path turns are a combination of coordinated roll and yaw rotations that result from pilot initiated motion of both the *aileron-* (the hinged surface at the edge of an airplane wing) and rudder-control surfaces. Figure 2.3 shows the three rotational motions and the corresponding axes as they apply to a quadcopter. The quadcopter diagrammed in the figure is in an *X* configuration, which is discussed in the next section. However, it makes no difference how a quadcopter is configured; the pitch, roll, and yaw rotational motions will always be the same for each axis.

Basic Quadcopter Configurations

The basic quadcopter is simply a center platform from which booms are extended. Motors with propellers are attached at the end of each boom. A variety of configurations exist based upon this basic form. Some of the most common are shown in Figure 2.4.

The configuration shown at the top left of the figure is the *plus* configuration, while the top center is the *X* configuration. The *X* configuration is the type used in the Elev-8. The only difference between the *plus* and *X* configurations is the forward direction designation. The forward direction is always aligned with a boom for the *plus*, while it is centered between two booms for the *X* configuration. It is imperative that the actual quadcopter configuration be input into the flight-control board, or else it will not be able to properly control the copter.

Name	Axis	Description
Pitch	Lateral	Rotation around the lateral axis that results in a climb or descent
Roll	Longitudinal	Rotation around the longitudinal axis that results in straight line roll but no turn to either side
Yaw	Vertical	Rotation around the vertical axis that results in a turn to the left or right

TABLE 2.2   Aerodynamic Rotational Motions

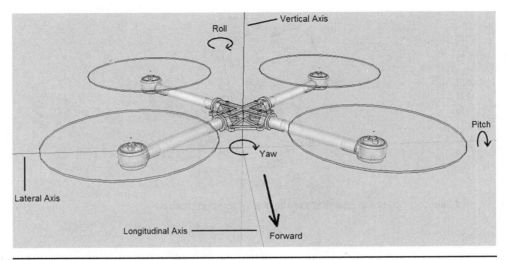

FIGURE 2.3   Quadcopter principal axes and respective rotational motions.

In Figure 2.4, all the motor positions have the rotation designations for *clockwise* (CW) or *counterclockwise* (CCW). I will ignore the ones with both for now. When viewed from above, the CW and CCW rotations alternate, which ensures that there is a net zero torque on the quadcopter, and thus, it will not yaw when all motor speeds are equal. A strong yaw rotation would result if all the motors rotated in the same direction, because of Newton's third law of motion: *To every action there is always an equal and opposite reaction*. It is possible to start yaw rotation by varying the speeds of the CW and CCW motors separately.

Different propellers are used on the CW and CCW motors. Figure 2.5 shows one of the propellers that is designed to turn CCW. The propeller is part of the Elev-8 kit and is designated as a 10 × 4.7 Slo-Flyer Pusher, model LP 10047 SFP. The 10 in the designation refers to the propeller diameter in inches, while the 4.7 refers to the number of inches that the propeller advances into the air per revolution. This number is also referred to as the

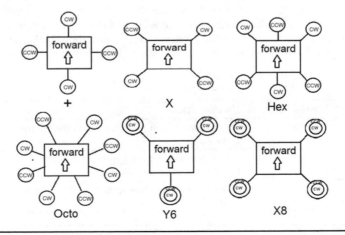

FIGURE 2.4   Basic quadcopter configurations.

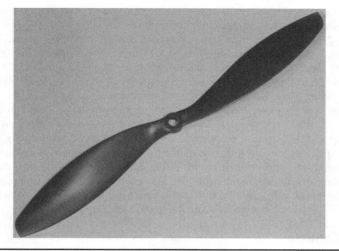

Figure 2.5   CCW Elev-8 propeller.

propeller's pitch, not to be confused with the aerodynamic pitch describing an airplane's attitude. A large propeller pitch number means it takes a large "bite" out of the air for every rotation. Conversely, a small pitch number means smaller "bites" are taken by the propeller. Large pitch also means more torque is required from the driving motor, which translates to more power required for that motor. The Slo-Flyer description refers to the propeller being designed for relatively fewer *revolutions per minute* (r/min) than higher speed propellers. Slo-Flyer propellers have a top r/min limit of about 7000 r/min, while high-speed propellers can exceed 15,000 r/min. High-speed propellers often have a small pitch because it would be impractical to provide the power needed to drive a large-pitch propeller at such high speeds for any useful length of operating time.

The CW-turning propeller has the designation 10 × 4.7 Slo-Flyer, model LP 10047 and is shown in Figure 2.6. The only difference between the two propellers is the pitch angle, which

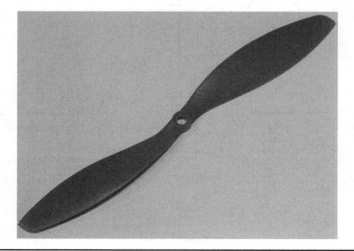

Figure 2.6   CW Elev-8 propeller.

is why it is critical to ensure that you mount the correct propeller on a motor whose rotation matches the propeller's maximum r/min rating.

Sharp-eyed readers may have spotted a copter with three booms on the bottom row, center of Figure 2.4. Naturally, this would indicate an unbalanced torque arrangement; however, the "Y6" has a clever trick to counteract the odd number of booms. At the end of each boom is one motor that drives two propellers, one at the top and one at the bottom. The top propeller turns CW, while the bottom propeller turns CCW, thus cancelling the top propeller's torque effect. Another more complex approach is to have two motors mounted at the end of each boom: one driving the top propeller and the other driving the bottom one. Either approach enables a multirotor copter to have an odd number of booms if so desired.

The quadcopter configuration on the bottom left is known as an "X8" because it has two propellers at the end of each boom. Either one motor drives both propellers, or there are two motors, one to drive each propeller. Having twice the number of propellers increases the available thrust substantially, but at the expense of requiring a lot more power for every motor as compared to a regular four-bladed quadcopter.

Flight Controls

It would be useful now to describe how normal airplane flight controls function before describing how the quadcopter flight path is controlled. The reason is simply that the *radio-controlled* (R/C) system is set for controlling an airplane, not for controlling the quadcopter, and it is important for you to know the "translation" that takes place when you input a control command. Figure 2.7 shows the external control surfaces that can change the pitch, roll, and yaw of an airplane based upon pilot commanded control movements.

Figure 2.8 shows the interior of a modern Cessna 172S equipped with a Garmin G1000 avionics suite, commonly referred to as a "glass" cockpit. For purposes of this discussion, your attention should be focused on the yoke, rudder pedals, and throttle that are pointed out in the figure. The pilot, who is normally in the left seat, changes the pitch attitude by

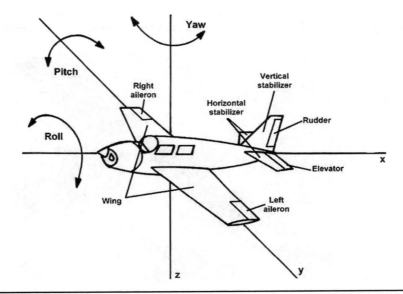

FIGURE 2.7 Airplane control surfaces.

FIGURE 2.8   Cessna 172S cockpit.

pulling on the yoke to climb and pushing on it to descend. Throttle changes are often needed as part of the climb or descent maneuvers. The external, elevator-control surfaces, shown in Figure 2.7, are the ones involved with climbing and descending.

To make coordinated turns, turn the yoke left or right, and simultaneously press on the appropriate rudder pedal that matches the turn direction. Using the rudder alone will turn the aircraft, but it would result in an unpleasant turn in which the aircraft would either slip or skid throughout the turn. The external, aileron- and rudder-control surfaces, shown in Figure 2.7, are the ones involved with turning. Turning only the yoke will roll the airplane around the longitudinal axis without changing its direction of travel. The ailerons are used solely as the external control surfaces.

Quadcopter Controls

Now that the basic airplane controls have been discussed, we can begin the discussion of the quadcopter controls. The quadcopter is controlled as if it were a normal R/C guided airplane. The difference in control happens when the quadcopter's flight-control board intercepts the normal flight-control commands and translates them into appropriate motor speed control signals. That is all that can be controlled on a quadcopter, which lacks the wings, ailerons, rudder, and flaps found on a normal aircraft. Figure 2.9, taken from the Spektrum DX-8 user's manual, shows the transmitter I used to control the Elev-8.

The stick on the left controls both the throttle and the rudder, while the stick on the right controls the ailerons and the elevator. Pushing the left stick forward and back increases or decreases the motor speed to all motors, respectively. Increasing all motor speeds simultaneously will send the quadcopter into a vertical flight path that is equivalent to a climb in a normal aircraft. Obviously, an equal simultaneous reduction in power causes it to descend. A somewhat more interesting control action happens when the right stick is moved left or right, a movement that ordinarily controls the elevator of a normal airplane. Changing

TRANSMITTER IDENTIFICATION MODE 2

Note: to change transmitter modes see page 39

⚠ **WARNING: ENSURE FUTURE ANTENNA SAFETY** Do not attempt to use the antenna to bea any weight, pick up the transmitter by the antenna or alter the antenna in any way. If the transmitter antenna or related components become damaged the output strength can be severely impeded which could lead to a crash, injury, and property damage.

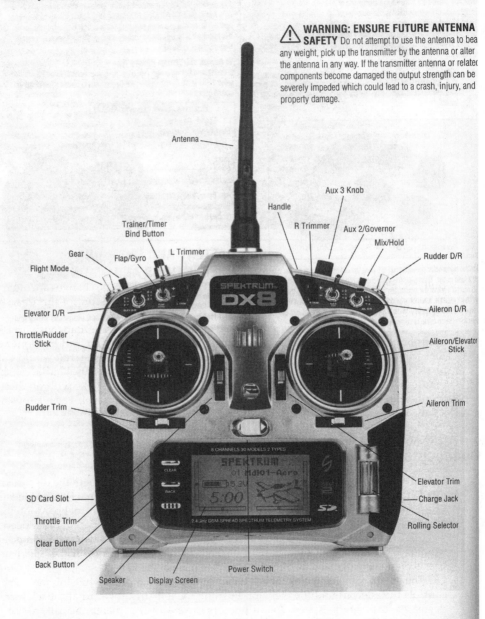

FIGURE 2.9   Spektrum DX-8 R/C transmitter.

the elevator changes the pitch attitude in a normal aircraft. The quadcopter change in pitch happens by altering the speed of both motors forward of the CG. Reducing the speed should pitch the quadcopter forward, and it will proceed in a forward direction. However, it is not simply a matter of changing motor speeds because the altitude at which the quadcopter is operating should not change as a result of the pitch command. All the motor speeds must change, both to maintain altitude and to effect a pitch rotation. The following set of equations should help to clarify all of the quadcopter operations. Figure 2.10, which is a modified version of Figure 2.3, shows all the motors with matching equation and rotation identifiers.

MP1 = Motor speed for the left front motor

MP2 = Motor speed for the right front motor

MP3 = Motor speed for the left rear motor

MP4 = Motor speed for the right rear motor

T = Throttle setting

Straight up or down vertical flight:

$$MP1 = MP2 = MP3 = MP4 = T$$

Pitch change in a hover state:

The motor speeds for MP1 and MP2 must be changed in order to pitch the quadcopter about the lateral axis. However, only changing these two motor speeds will upset the altitude that is established. Therefore, the flight-control board computes an offset speed that it subtracts from both forward motors, while it adds the same offset to both rear motors, thus allowing for a pitch change but not changing the overall throttle setting. This ensures that the quadcopter does not change altitude.

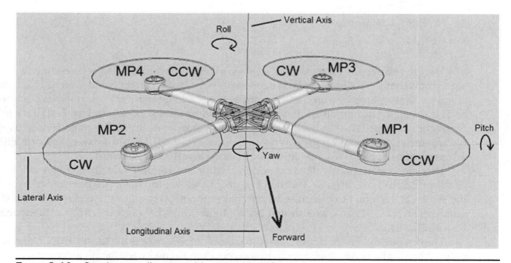

Figure 2.10 Quadcopter diagram with motor identifiers.

$$MP1 = T - \text{offset}$$
$$MP2 = T - \text{offset}$$
$$MP3 = T + \text{offset}$$
$$MP4 = T + \text{offset}$$

You should also realize that you could increase the throttle while maintaining a pitch change, which would result in putting the quadcopter into a normal ascent versus a straight vertical climb.

Yaw change in a hover state:

Placing the quadcopter in a yaw without changing altitude is similar to a pitch change except that the lower speed changes are applied to motors controlling the desired opposite yaw direction. This means that a desired CCW yaw would have an offset subtracted from both CW motors and the same offset added to both CCW motors in order to maintain altitude.

$$MP1 = T + \text{offset}$$
$$MP2 = T - \text{offset}$$
$$MP3 = T - \text{offset}$$
$$MP4 = T + \text{offset}$$

Roll change in a hover state:

Rolling the quadcopter is a matter of increasing the speed of both motors on the side opposite to the desired roll direction and simultaneously decreasing the speed of both motors on the other side. Below are the equations for a roll to the left.

$$MP1 = T + \text{offset}$$
$$MP2 = T - \text{offset}$$
$$MP3 = T + \text{offset}$$
$$MP4 = T - \text{offset}$$

The preceding set of equations is very straightforward and is representative of the algorithms that the flight-control board implements. However, the quadcopter flight control is not quite that simple. Automated control of a quadcopter aircraft means that there must be at least one sensor involved that reports the condition and position of the craft back to the flight-control board so that the repositioning can stop as desired. The main sensor used in the HoverflyOPEN board is the Invensense model ITG-3200, MEMS 3-axis gyroscope. Figure 2.11 is a photo of the gyroscope mounted on the HoverflyOPEN board.

This sensor can rapidly detect minute variations in angular velocity changes in all three of the principal axes discussed earlier. Figure 2.12 shows the three predetermined axes that the sensor is designed to measure, which makes it critical to align these axes with the three quadcopter axes. The $+Y$ axis shown on the figure must be aligned with the quadcopter's forward direction.

The dot printed on the upper left hand corner of the sensor is the key to proper alignment on the board. The board itself must also be properly aligned with the quadcopter's forward

Figure 2.11   Invensense ITG-3200 gyroscope.

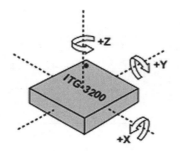

Figure 2.12   ITG-3200 sensor axes.

direction. Failure to properly align the board means the gyroscope cannot accurately measure the appropriate angular velocities, thus making quadcopter control questionable.

Raw data on each axis is sent in serial format from the gyroscope sensor to the main processor on the flight-control board at a very fast rate. This main processor is the Parallax Propeller chip, which will be thoroughly discussed in Chapter 4. What should be noted now is that a great deal of information is extracted from the raw data by some very involved and complex calculations in order to generate the appropriate motor-control speed commands that reflect what the user wants to do with the quadcopter. There is also a good deal of ongoing real-time filtering to ensure that only the relevant user commands are being followed and are not being disturbed by noise.

PID Control

PID is an acronym for *proportional integral derivative* and is used in almost all quadcopter control systems. The theory behind PID is relatively simple to understand and begins with the block diagram shown in Figure 2.13.

All control systems have process variables that are required to be at a specific value. For example, the thermostat is part of the very familiar home-heating (and maybe cooling) system. The room temperature is the process variable in such a system. We could set a temperature on the thermostat, and if that value was higher than the actual room temperature, the thermostat would direct the furnace to heat the room by using the available heating system (such as hot air or hot water). The system would continue to provide heat to the room until the new temperature was reached. As the room cooled naturally from heat losses through windows and open doors, the room temperature would drop below the set point and cause the control process to repeat. The heat losses are called system disturbances and are the reason why the thermostatic-control system is needed. All real-world systems have their own disturbances, and thus, need a control system to maintain the balance, equilibrium, or set point. Table 2.3 relates the above system operation to the Figure 2.13 elements.

The thermostatically controlled room heating system is an example of a closed-loop control system. A sensor continuously reads the room temperature and provides this to the controller, which already has a set point. The difference between the real-time sensor reading and the set point is the error signal used by the controller to actuate the system or plant, such that the error drives toward a zero difference. Sometimes there is an offset value permitted between the sensor value and the set point when it is not realistic to obtain a zero error or the system functions require an offset.

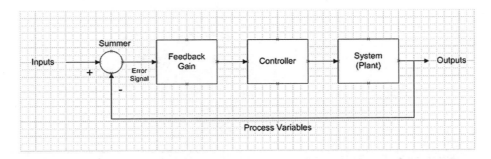

FIGURE 2.13 PID block diagram.

System Component	Description
Set point	The temperature set on the thermostat
Process variable	Room temperature
Error signal	Set point minus process variable
Output	Abstraction for the room heating process
Feedback gain	Normally zero for this type of control system
Controller	Varies with thermostat type—might be a microcontroller; old fashion type uses bimetallic electrical contacts
Plant	Furnace and all its related components

TABLE 2.3   System Components Related to PID Block Diagram

Control systems are normally designed to minimize disturbance effects. Several key parameters that characterize control system performance need to be defined. These are:

- Rise time
- Percent overshoot
- Settling time
- Steady-state error

Figure 2.14 shows a typical PID response graph with the previously mentioned key parameters annotated within the figure. This graph represents the process variable response

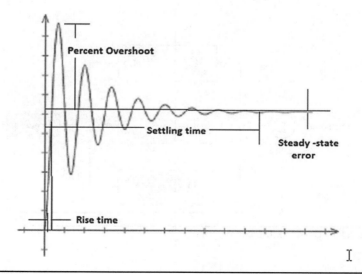

FIGURE 2.14   PID response graph.

to a step input applied to the control system. Time is usually the *X* axis, while the *Y* axis will normally be the process variable units, such as the temperature degrees in our thermostat example. The following definitions are commonly but not universally accepted in the control industry:

- *Rise time*—The time to go from 5% to 95% after the step is applied.
- *Percent overshoot*—The peak value of the response expressed as a percentage of the steady-state value.
- *Settling time*—Time to settle to within a certain percentage of steady state. Often chosen at 5% but not guaranteed.
- *Steady-state error*—The actual output versus the ideal output.

Several other performance parameters are also used to help characterize control systems. These are:

- *Deadtime*—A measure of the time delay between a process variable change and system recognition of that change.
- *Loop cycle*—Time between calls to the control system algorithm.

Both of these parameters will have a significant impact on a quadcopter control system. Minimizing dead time and loop-cycle timing is critical to optimizing the control algorithms. Careful optimization of the source code and incorporation of the assembly language routines, where necessary, will go a long way toward minimizing both of these parameters.

PID Theory

It is time to examine the PID theory now that the system configuration and definitions have been covered. Figure 2.15[1] shows the classic PID block diagram. Each section will be discussed separately.

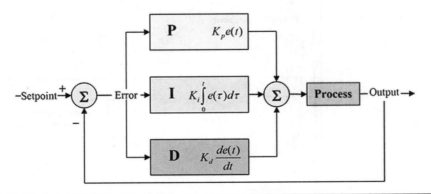

Figure 2.15   Classic PID block diagram.

---

[1]Wikipedia PID block diagram.

P Block

The P or proportional block depends only upon the difference between the set point and the process variable. The P block math equation is

$$K_p e(t)$$

where K_p is the block gain and $e(t)$ is the error signal as a function of time. This is a simple linear system response. For example:

> Given that the error signal at time t_0 is 5 and K_p is 10, then the output from this block is 50. You must be careful in setting the value for K_p. Too high a value will make the system unstable and fall into an oscillation state that would be very bad for system operations. A procedure for setting K_p along with the other block constants will follow this section.

I Block

The I or integral block depends upon a summation of the error signal over time. The I block math equation is

$$K_i \int^t e(\tau)d\tau$$

where K_i is the block gain and $\int^t e(\tau)d\tau$ is an integral equation of the error signal as a function of time.

The integral equation is additive, which allows the output to steadily increase over time unless the error signal is zero or there are compensating negative error values. The net effect of the integral block is to drive the steady-state value to zero overall during a time period. The nominal value for the I block gain is usually very small, which you might expect since the term acts over a long time period.

There is one issue that arises with this block: the integral term could temporarily increase to a level that saturates the plant block without driving the steady-state value toward zero. This is called *integral windup*. Windup, while a potential issue, is not expected to happen in the typical quadcopter control system that is discussed in this book.

D Block

The D or derivative block depends upon rapid increases in the process variable to drive the error signal to zero. The D block math equation is

$$K_d \, de(t)/dt$$

where K_d is the block gain and $de(t)/dt$ is a derivative equation of the error signal as a function of time.

This block gain must be chosen carefully to allow the system to respond to rapid process changes, yet not to over respond to noise added to the feedback loop. The practical tradeoff is to set a low value for the K_d gain and a small derivative time Δt that approximates $de(t)/dt$.

Tuning

Tuning is the process of determining useful gain values to use with the PID algorithm. Two methods will be discussed:

1. Trial and Error Method
2. Ziegler-Nichols Method

Trial and Error Method

Do not be deterred by this method's name because there is a definitive approach that is followed in this method. The steps are:

1. Set the K_i and K_d block gains to zero.
2. Increase the K_p block gain until the system becomes unstable, as determined by observing system oscillation.
3. Increase the K_i block gain to stop the oscillation induced in step 2.
4. Increase the K_d block gain to improve the system time response to an acceptable value.

By following the above steps, a reasonable set of block gains should be able to be set. These values should be tweaked to reach acceptable performances in actual operation. For instance, it may be observed that the K_d block gain is set too high because the system is too reactive to noise.

Ziegler-Nichols Method

This tuning method is similar to the Trial and Error method in that the first two steps are identical.

1. Set the K_i and K_d block gains to zero.
2. Increase the K_p block gain until the system becomes unstable as determined by observing system oscillation.
3. Note the K_p block gain at which oscillation starts. This will now be called critical gain or K_c. Also note the oscillation period. This will be called P_c.
4. Adjust all block gains per Table 2.4.

As you can clearly see from Table 2.4, control systems do not always have to contain all three of the PID blocks. Sometimes only the P block is needed, as we saw in our thermostat example.

Real-world PID control systems often contain an auto tuning capability in which the block gains and response times are both detected and set to optimize system operation. *LabVIEW* (LV) discussed in the next section contains a PID module that has an auto-tune capability as well as a manual tuning capability.

LabVIEW PID Simulation

I will be using the LV 2012 student edition for my simulation platform. It is the latest student version available at the time of this writing. Be aware that commercial, non-educational versions of LV cost about $1200. It is not a cheap program. However, a student version is

System Type	Block Gain	Integral Time	Derivative Time
P	0.5 K_c	n/a	n/a
PI	0.45 K_c	$P_c/1.2$	n/a
PID	0.60 K_c	0.5 P_c	$P_c/8$

TABLE 2.4   Ziegler-Nichols Block Gains

available that is very reasonable in cost and even comes with an Arduino microprocessor. This package is available from Sparkfun.com. The student version is not crippled in any way; users are simply not permitted to develop commercial (meaning for sale) products incorporating LV without purchasing a retail LV license.

Every program in LabVIEW (LV) is called a *virtual instrument* or VI. This naming system is from the very earliest days (1986) when LV was first created to control electronic test equipment. LV has evolved to far more than instrumentation control, but it still pays homage to its beginnings with the file extension .vi. Clicking on the Blank VI icon will create the next screen, which is shown in Figure 2.16.

A careful look at this figure should reveal two screens, one named *Front Panel* and the other *Block Diagram*. LV creates these two views for each VI. The *Front Panel* is the user interactive screen or GUI, while the *Block Diagram* houses the code. All LV programs consist of Function Blocks interconnected with wires—the kind of wires that carry data not electricity. Functions are selected off a series of palettes, such as the one shown in Figure 2.17. Palettes are selected from the View drop-down menu. The palette selected is the Controls Palette with the Express controls displayed.

I did not create a PID VI myself but instead downloaded one named Single Axis Quadcopter.vi developed by National Instrument application engineers to demonstrate LV's ability to simulate a portion of a quadcopter's functions. This VI is available at https:// decibel.ni.com/content/docs/DOC-22670. There is one limitation that you should be aware of before running this VI. It depends upon two sub-VIs that are part of the LabVIEW 2012 PID and Fuzzy Logic Toolkit. This software toolkit is an expensive add-on to LV; however, you are able to download it from www.ni.com and install it for a limited time in an evaluation mode. This was how I was able to run the single axis simulation. Note that there are only two

Figure 2.16   Blank VI screenshot.

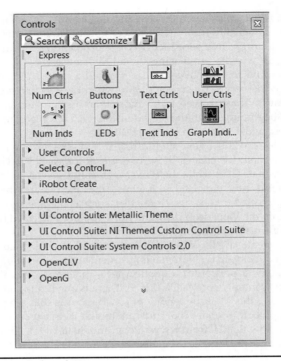

FIGURE 2.17   Express controls palette.

motors, MP2 and MP4, being simulated in this VI because it is a single-axis demonstration. Figure 2.18 is a screenshot of the VI running.

I set the quadcopter setpoint to +10° for this simulation. After running the simulation for several seconds, the virtual quadcopter settles into a 10° bank, as you can see from both the diagram in the top center and the indicator to the right of the quadcopter diagram. The time-response chart in the lower right shows that the quadcopter smoothly went to the commanded bank in about one second without any overshoot. This is precisely the type of control behavior that you should want from a quadcopter flight-control board. Also, notice that the three PID parameters, K_p, K_i, and K_d may be set to experiment with different values in order to test how they affect overall performance.

I also ran another simulation, which is shown in Figure 2.19 but this time I increased the throttle to 70% and increased the angle to 25°.

Take a look at the waveform chart in the lower left corner. MP2 is running at 74% while MP4 is at 70%, which causes the 25° pitchover to happen. Compare this waveform to that shown in Figure 2.18, where MP2 is running at approximately 52% and MP4 at 50%, and which causes the shallower bank angle of 10°.

A much more challenging simulation that you can try is the Untethered Quadcopter.vi from the same group that created the first VI discussed. Figure 2.20 is a screen shot of the program running.

Here the challenge is to try to keep the quadcopter figure in the visible flying box by using a combination of throttle and setpoint angle. Believe me, it is not easy. In fact, this

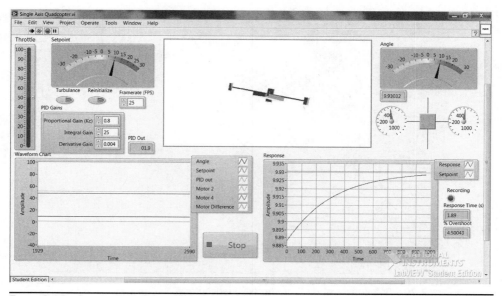

FIGURE 2.18 Screenshot of the single-axis Quadcopter VI running at 50% throttle and 10° angle.

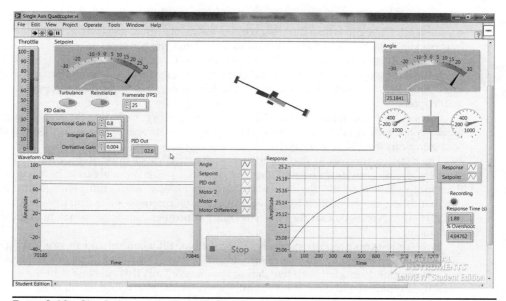

FIGURE 2.19 Simulation screenshot with throttle at 70% and angle at 25°.

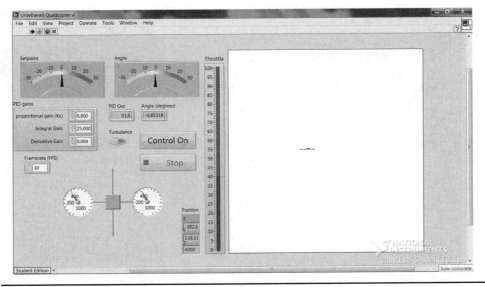

FIGURE 2.20   Screenshot of the Untethered Quadcopter.vi.

simulation actually provides a bit of insight into the skills needed to pilot the real quadcopter, although I believe the simulation is more sensitive than the real item.

Summary

Because it is important to have a good foundation for studying the particulars of quadcopter aerodynamics, I began with a discussion of the basics of normal airplane aerodynamics. The four principal flight forces were introduced along with the concepts of weight and balance and the closely allied concept of center of gravity (CG).

Basic flight axes were than described along with the corresponding rotations about these axes: pitch, roll, and yaw. A diagram showed how these axes and rotations applied to the quadcopter.

Some of the basic quadcopter configurations were shown, including the X configuration, which is the type used in the Elev-8. The two types of propellers used in this and other configurations were also described.

The detailed discussion that came next focused on quadcopter flight control because it differs significantly from normal airplane control. I presented a series of basic equations that encapsulate how the flight-control system translates normal airplane flight-control commands to the type needed by a quadcopter. A brief discussion was also presented on the MEMS 3-axis gyroscope that the HoverflyOPEN controller uses to sense actual quadcopter movement.

Two sections on proportional-integral-derivative (PID) control and theory were given, as that is the key technology to ensure smooth and positive control for quadcopter operations.

The chapter concluded with two demonstrations of LabVIEW, the first running a single-axis PID simulation and the second an untethered quadcopter flying simulation.

As a reward for making it through this chapter, I will now put aside the heavy theory and go on to show you how to build your own Elev-8 quadcopter.

CHAPTER 3

Building the Elev-8

Introduction

This is a long chapter that will take you through the complete process of building an Elev-8 by using the standard kit parts and a few additional components that are not part of the kit. Most of the instructions, figures, and assembly drawings in this chapter were created by Parallax for the *Information and Assembly Guide* that accompanies their Elev-8 Quadcopter Kit (#80000) and are being presented with Parallax's kind permission. (It made no sense for me to create a new set of instructions when a perfectly acceptable and useful set was already in existence.) I did take the liberty of adding some additional figures and written instructions where I thought it would be useful. I also added my own instructions for some highly useful components that will make your build a bit easier. The entire build should take anywhere from 10 to 12 hours, provided you have all the requisite parts, tools, and a functional work space. I will discuss the tools and supplemental materials, but first it is important to review some safety tips.

Safety

The following bulleted safety items were taken from the Parallax document cited above:

- *WARNING: CUTTING HAZARD. Rotating ELEV-8 quadcopter blades can cut skin and underlying tissues. Stay away from a powered ELEV-8 quadcopter and never become complacent during operation.*
- *WARNING: ENTANGLEMENT HAZARD. Secure long hair and loose clothing or jewelry when building, testing, and operating your ELEV-8 quadcopter to avoid entanglement with motors.*
- *WARNING: EYE HAZARD. Always wear eye protection when assembling, soldering, operating, or repairing your ELEV-8 quadcopter.*
- *Inform yourself of and follow all current federal, state, and local laws regarding the use of hobby RC aircraft in the area where you plan to operate your ELEV-8 quadcopter. Review the FAA's rules in entirety—you are responsible for following them.*
- *This kit is not for beginners. Advanced mechanical skill is required for building and flying an ELEV-8 quadcopter. RC aircraft experience is highly recommended.*

- *Follow the instructions carefully; incorrect assembly of your ELEV-8 quadcopter could cause risk of catastrophic equipment failure, personal injury to you or others, and property damage.*
- *Perform initial* electronic speed controller *(ESC) programming before installing the propeller blades. Remove propeller blades before reprogramming the ESCs.*
- *Establish and test the radio link between the RC controller and RC receiver before installing the propeller blades. Remove propeller blades before testing a different controller.*
- *Always disconnect the battery when not in use.*
- *Store your ELEV-8 quadcopter and its radio controller out of reach of children, pets, and those who do not know how to use them safely.*
- *Only operate your ELEV-8 quadcopter in an area with no children, unsecured pets, or livestock, which can be harmed by contact with rotating blades. For example, children and dogs may try to jump and catch a flying quadcopter, or may run to investigate one that has just landed.*
- *Only operate your ELEV-8 quadcopter outdoors and away from crowded areas. All observers should stand a safe distance* **behind** *the operator.*
- *Only operate your ELEV-8 quadcopter in an environment where you can maintain unobstructed visual contact with it. Do not operate at night, or where there is fog, smoke, or dust that could limit visibility.*
- *Keep your ELEV-8 quadcopter dry! Do not submerge your ELEV-8 quadcopter or operate it in rainy or damp conditions. Beware of sprinklers and of landing in wet vegetation.*
- *Check the wind speed before flying your ELEV-8 quadcopter. Even a light breeze can make flying difficult for beginners. No one should fly in high winds.*

Most of the above safety items listed are common sense. I did not list several items from the original document, since I believe that they were based more on a legal perspective with regard to the company selling a product that might cause damage and/or injury if not used properly. A complete written copy of the Parallax Assembly instructions is included in every Elev-8 kit sold. Please refer to that document to read the original version.

Tools and Additional Materials

The following is a recommended tool list that you will need to build the Elev-8:

- Soldering iron and rosin-core solder (acid-free solder flux optional)
- Component clamp stand
- #1 Phillips screwdriver
- ¼-in (0.6-cm) wrench, box-end or socket
- $^{11}/_{32}$-in (0.9-cm) wrench or nut driver
- Wire strippers/cutters (12-16 AWG)
- Scissors
- Needle-nose pliers
- Diagonal cutter
- Ruler or measuring tape
- Heat gun

I would recommend a soldering iron or solder station that is temperature controlled, as you will need quite a bit of heat to solder some large connectors and multiple wire assemblies. Other, smaller connectors will not require that much heat energy to melt the solder. Also, I recommend that you buy the best quality rosin-core solder that you can find. It is well worth the investment. Cold solder joints are the bane of any builder, and they are hard to locate. The key to quality soldering work is to have good soldering technique, keep the soldering iron tip clean, and to use the highest-quality solder available. Figure 3.1 shows the essence of good soldering technique. It is vital that the solder joint be hot enough for solder to flow easily. It takes practice to apply just the right amount of solder; too little could result in a cold solder joint, and too much could lead to a short between closely spaced components.

Another issue regarding a good solder joint is the use of lead-free solder. Now, please don't get down on me. I am all about maintaining a healthful environment, but the elimination of lead from solder often produces poor solder joints unless some extra precautions are taken. The simplest, and probably, the best approach is to apply a high-quality, acid-free solder flux to the joint prior to heating the joint with the iron. This will allow the lead-free solder to flow more freely and produce a better-soldered connection. Again, it takes practice to perfect soldering techniques.

I have one final thought that relates to solder joints as well as other types of electrical connections. A long-running anecdotal observation contends that 90 percent of all electrical/ electronic malfunctions are related to connection malfunctions. This claim makes a lot of sense when you think about it. We live in an oxygen rich atmosphere, and oxygen is a great reduction agent: it wants to oxidize every element it can possibly chemically combine with. Metal oxides are reasonably good insulators because some of their free electrons have been taken up by oxygen molecules. This leads to higher and higher resistance being built up in a connection, which will eventually cause a failure. Of course, current flowing through a resistance produces heat, which in turn can cause a fire if the currents are sufficiently high. So, what is the solution? One expensive solution is to gold plate electrical contact surfaces. Gold doesn't oxidize and is not subject to this type of failure. It is, of course, very expensive and not practical for large-scale connectors. For the type of projects that I work on, I can ensure only that solder joints are sound from both a mechanical and electrical perspective. I also inspect electrical connections for oxidation and foreign matter, and take appropriate action to replace or repair a damaged component.

Bill of Material

The *bill of material* (BOM) is reproduced below in Figures 3.2 to 3.5. It is highly recommended that you cross-check all the material in the kit against the BOM and contact Parallax if anything is missing. I have always found them very prompt and attentive to requests from builders, in instances where parts are missing or otherwise incorrect. All of the kit contents

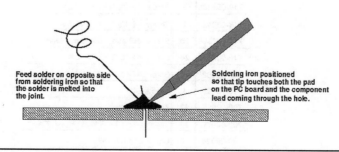

Feed solder on opposite side from soldering iron so that the solder is melted into the joint.

Soldering iron positioned so that tip touches both the pad on the PC board and the component lead coming through the hole.

FIGURE 3.1 Good soldering technique.

ELEV-8 Quadcopter Kit Contents (#80000)		
Part #	Qty	Description
31500	1	Hoverfly OPEN Board
700-10003	1	Safety glasses
80050	1	ELEV-8 Airframe Kit
80060	1	ELEV-8 Hardware Kit
80070	1	ELEV-8 Electronics Kit
85000	1	ESC Programming Card*

FIGURE 3.2   High-level BOM.

ELEV-8 Airframe Kit Contents (#80050)		
Part #	Qty	Description
730-00060	4	ELEV-8 Boom (Black)
721-80010	1	ELEV-8 Control Board Top Plate
721-80007	4	ELEV-8 Landing Gear
721-80005	4	ELEV-8 Motor Mount Top/Bottom
721-80003	2	ELEV-8 Quad Chassis
721-80002	1	ELEV-8 Control Board Mount Plate

FIGURE 3.3   Airframe BOM.

ELEV-8 Hardware Kit (#80060)		
Part #	Qty	Description
900-00021	2	Nylon Strap, Black
725-00067	1	1.5 mm Hex Key
720-28001	1	Light Pipe
713-00051	4	Spacer,#4x1/2",NY
713-00043	16	Standoff, #4-40, 5/8", Nylon
713-00025	4	Standoff, #4-40,1 1/4", Nylon
712-00004	4	Washer, #6, 3/8" OD, Zinc
710-00100	8	Screw, #4-40, 1/4", PH, Black
710-00042	4	Screw, #4-40, 1-1/4", PH, SS
710-00039	16	Screw, 3x6 mm,0.5 thread, black
710-00036	24	Screw, #4-40, 3/8", PH, SS
710-00002	16	Screw, #4-40, 1",PH,SS
700-00106	1	Loctite 242
700-00093	12	Zip Tie, 4" Black
700-00059	16	Internal-tooth Lock Washer, #4, Zinc
700-00024	4	Locknut #4-40, 1/4"

FIGURE 3.4   Hardware BOM.

ELEV-8 Electronics Kit (#80070)		
Part #	Qty	Description
800-00080	5	3-wire ext.,22AWG,F/F,8"
800-00039	6"	1/2" Heat Shrink Tubing (black)
800-00036	24"	3/4" Heat-shrink Tubing (clear)
800-00023	27"	3/16" Heat-shrink Tubing (black)
800-00022	2	16" Black Pluggable Jumper-Male
800-00021	2	16" Red Pluggable Jumper-Male
750-90002	4	Brushless 1000Kv Motor
750-90000	4	Gemfan 30A ESC Speed Controller
750-00059	18"	12AWG Red Wire
750-00058	1.5	12AWG Black Wire
750-00056	15	16AWG Red Wire
721-80001	2	10x4.7 Slow Flyer CW 1045R Blade
721-80000	2	10x4.7 Slow Flyer CCW 1045 Blade
452-00088	1	EC3 Plugs(10 pairs/sets)
450-00050	2	Extra Gold Bullet Conn 10-Pak
350-00045	8"	White LED Tape
350-00044	8"	Red LED Tape
120-00007	8"	Red/White Checkered Sticker
120-00006	8"	Black/White Checkered Sticker
800-00080	5	3-wire ext.,22AWG,F/F,8"

FIGURE 3.5   Electronics BOM.

are shown in Figure 3.6, including the two safety glasses that are thoughtfully included by Parallax for builder safety.

I highly recommend that you obtain two plastic containers that can be set up to store most of the kit pieces. I would also recommend that after opening each plastic bag in the kit, you place the pieces directly in a separate container and label each one for your convenience. Use the Elev-8 V1.2 Assembly Drawings, found online at www.mhprofessional.com/quadcopter, as a guide to matching the pieces to the part number or assembly step. Figure 3.7 shows one of the plastic containers holding all the hardware pieces and a few of the electronic pieces.

One word of caution: it is somewhat difficult to differentiate between the ¼-in 4-40 black screws and the 3-mm × 6-mm black screws. I would suggest separating the screws and then counting them. There are eight 4-40 screws, while there are sixteen 3-mm × 6-mm screws.

Additional Materials

You will need the following items to be able to fly your Elev-8:

- An R/C transmitter and receiver
- A three-cell 30C LiPo battery
- A LiPo battery charger

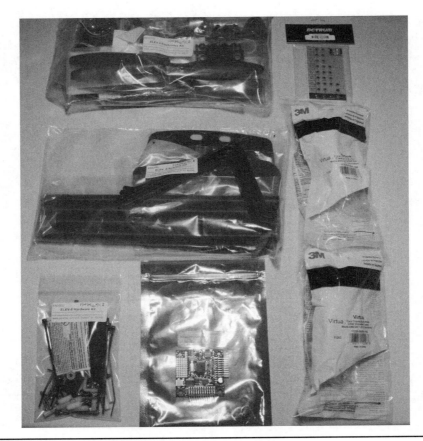

FIGURE 3.6   Elev-8 kit contents.

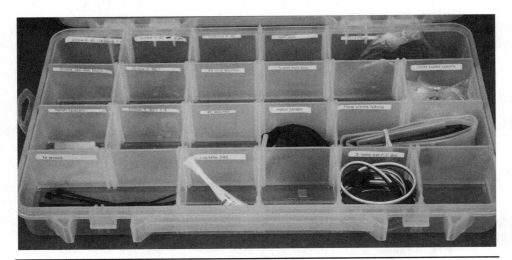

FIGURE 3.7   Plastic container holding all the hardware and some electronic pieces.

My recommendation is to purchase the Spektrum DX-8 R/C transmitter and the matching AR8000 receiver. Figure 3.8 shows a photo of the AR8000 receiver. You can find a photo of the DX-8 transmitter in Chapter 2, Figure 2-9.

The DX8/AR8000 combination is somewhat expensive; however, you will not be sorry to have purchased this set. The transmitter is excellent, and it comes with many features, some of which you will use immediately and others that will be available as you progress in your experience and training. The transmitter and receiver make eight control channels available, which is plenty to meet your immediate and future requirements. It is really not to your advantage to buy equipment that just meets your current needs. You would only be delaying a future purchase needed to keep up with your changing requirements.

The LiPo battery is a choice that is a much less challenging decision. The costs are quite reasonable for a variety of the batteries. My initial recommendation would be a three-cell LiPo battery with a 30C rating. The three cells are important because they set the overall voltage at 11.1 VDC (nominal). This is the minimum necessary to adequately power the motors and electronics onboard the Elev-8. The 30C is a capacity rating that will power the quadcopter for almost 20 flight minutes. It is always a tradeoff between weight and capacity, but I have found the 30C rating to be a sweet spot for this particular quadcopter. Figure 3.9 is a photo of the LiPo battery used in my Elev-8 build.

Purchasing a good quality LiPo battery charger is also required. LiPo batteries *must be charged using a charger designed for their particular chemistry.*

Warning: *Do not use an automotive type of charger with a LiPo battery. Attempting to do so will likely cause the battery to flame and produce a lot of toxic smoke.*

Quality LiPo chargers have a balancing circuit that ensures that each cell is appropriately charged at the level it requires. It would be prudent to spend a little extra money on a high-

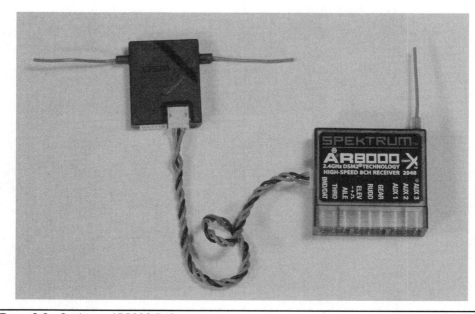

Figure 3.8 Spektrum AR8000 R/C receiver.

Figure 3.9 3S 30C LiPo battery.

quality LiPo charger. Figure 3.10 is a photo of the automatic LiPo battery charger that I use. It is a Thunder Power RC, model TP610C and is capable of recharging LiPo, NiCad, and lead-acid storage batteries.

Optional Additional Materials

Although not required, the following materials will either improve the build experience or add additional functionality to the quadcopter. Note that the Parallax Propeller Quickstart board is not included in this list because it is discussed in Chapter 4.

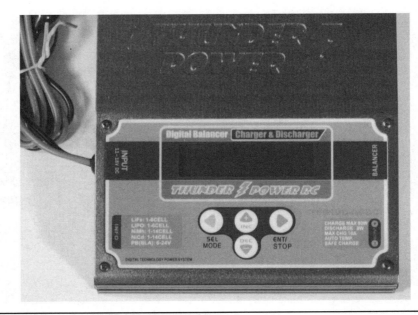

Figure 3.10 Automatic LiPo battery charger.

Build Related

- Power-distribution board: Quad
- Power-distribution cable harness
- Two additional EC3 connector kits (10 male/female pairs in each kit)
- 48 in of 22 AWG red solid core wire
- 48 in of 22 AWG black solid core wire

A power-distribution board is shown in Figure 3.11. Its purpose is to eliminate the cumbersome and awkward connections from the battery to the four *electronic speed controllers* (ESCs).

Without the power-distribution board, you must construct two somewhat bulky and clunky wire connection points between the battery terminals and the power wires that connect to the ESCs. Don't misunderstand me; you can do the latter, but this board just makes it so much easier and does so with a much cleaner appearance to boot. It makes use of wide electrical *printed circuit board* (PCB) traces that can easily handle the high currents created when operating the quadcopter. It also has plated vias (plated holes that connect to the underlying PCB traces) available where the *raw* battery supply can be accessed. This

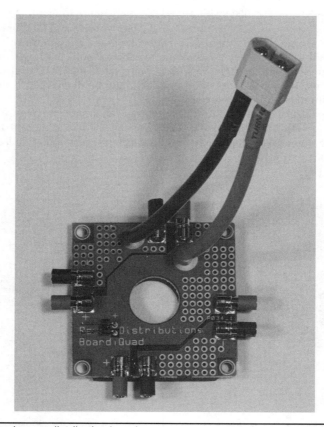

FIGURE 3.11   Quad power-distribution board.

makes it very easy to tap into the battery supply in order to power an accessory circuit or module. There is also a two-pin plug available for devices that use this common type of power supply connector.

A power-distribution cable harness is shown in Figure 3.12. It serves the same purpose as the quad power-distribution board but uses a premade harness that would be similar to the homemade one discussed above. Using the cable harness keeps the battery supply totally shielded; however, you would have no access to it without splicing into the harness.

I will be using a quad power-distribution board that I purchased from Hobbyking. The additional EC3 connectors will be used to connect the motors and extension wires. It is a process that is a bit more tedious; however, it makes for more sturdy and reliable connections.

The additional red and black 22 AWG wire will be needed to connect the LED lighting strips to the auxiliary Propeller-chip control board for light-control purposes. There will also be plenty of wire remaining for any circuits or modules that you might desire to add at a later date.

Functionally Related Material

- Spektrum TM1000 telemetry transmitter
- Spektrum telemetry brushless r/min sensor

The Spektrum TM1000 telemetry transmitter shown in Figure 3.13 sends data from the quadcopter back to the DX-8. It also comes with voltage and temperature sensors that can send real-time data back to the DX-8 for the voltage and temperature of a selected component to be monitored. I will use this module to monitor both the raw battery voltage and the battery temperature. The DX-8, in spite of being labeled an R/C transmitter, also contains a telemetry receiver. The data being sent back may be displayed on the DX-8 LCD screen, which is another reason to buy the DX-8.

The Spektrum telemetry brushless r/min sensor is shown in Figure 3.14. This sensor will be used to monitor the rotational speed of one of the Elev-8 motors. It will provide useful data when you are creating a new flight-control program and will also serve as a check on the real-time performance of the quadcopter.

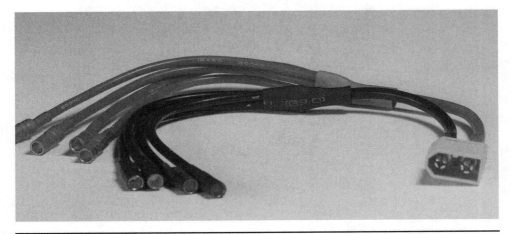

Figure 3.12 Power-distribution cable harness.

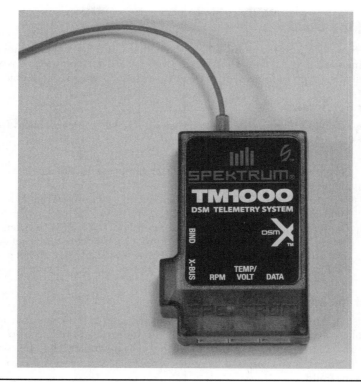

Figure 3.13   Spektrum 1000 telemetry transmitter.

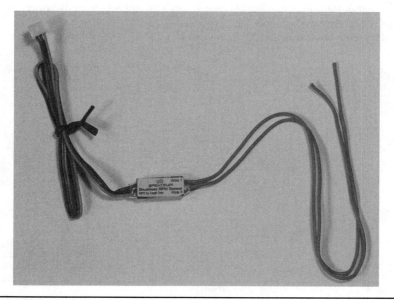

Figure 3.14   Spektrum telemetry brushless r/min sensor.

Beginning the Build

I have now come to the point where the actual build starts. Remember to have all the recommended tools, all the kit parts, and a well-lighted and comfortable work area available. These steps will go a long way in making the whole build experience enjoyable.

Motor Set Screws

Start by applying Blue Loctite to the motor set screws, to prevent them from coming loose during flight and causing equipment failure.

1. Locate the Blue Loctite 242, the four motors, and the small Allen wrench in the Elev-8 Hardware Kit.
2. Refer to Figure 3.15. Using the Allen wrench, carefully remove the motor set screws (item 2) from each motor (item 1).
3. The screws might be very tight; be careful not to break your Allen wrench.
4. For each motor, apply a small amount of Blue Loctite to the set-screw threads and carefully reinstall the screws. Seat each screw firmly but do not over tighten. Allow the Blue Loctite to set for 10 minutes. It fully cures in 24 hours.

Solder the Motor and ESC Connectors

In this step, you will solder EC3 connectors to each end of the long extension leads. You will also solder an EC3 connector to each of the motor leads. Finally, you will solder EC3 connectors to each of the ESC leads that connect to the motor. The EC3 connectors will give you the ability to switch around the wire connections when you check your motor direction later in the build.

Be sure to follow the specified genders for all of the leads. The general rule is if a lead is supplying power, then it will be a female connector. All connectors are eventually protected with shrink tubing; hence female connectors will not accidentally short out if disconnected from the mating male connector.

Please follow the instructions below.

1. Gather together your motors, the red 16 AWG wire, EC3 connectors, wire cutters, wire stripper, ruler, and soldering supplies.

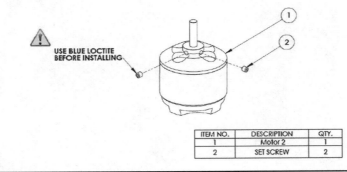

ITEM NO.	DESCRIPTION	QTY.
1	Motor 2	1
2	SET SCREW	2

FIGURE 3.15   Motor set-screw adjustment.

2. Using a ruler and wire cutters, measure and cut twelve lengths of the red 16 AWG wire; each one should be 12 in (30.5 cm) long. These wires will be referred to as extensions from this point on. Refer to Figure 3.16.

NOTE: *I have deviated a bit from the Elev-8 package instructions by having you add EC3 connectors between the wire extensions and the motors. I did this because it increases reliability at these critical connections and avoids a likely failure point if you inadvertently create a cold solder joint when connecting one wire to another. This is why I recommended that you purchase additional EC3 connectors over and above what is contained in the kit.*

3. Use wire strippers to remove the insulation from each extension and expose about ⅛ in (0.3 cm) of wire at each end. Pre-tin the exposed wires for easier soldering in the next step.
4. Next, solder a female EC3 connector at one end and a male EC3 connector at the other end. To solder an EC3 connector, insert the ⅛-in exposed tip of the wire into the cup end of the bullet connector, and fill the cup with solder but do not overfill. Figure 3.17 shows a soldered EC3 being held in a component clamp stand.
5. For this step, use all male EC3 connectors. Solder a male EC3 connector to the ends of all the motor leads.
6. If necessary, use wire strippers to expose ⅛ in (0.3 cm) of metal on the end of each speed controller's blue wire leads.
7. Solder a female EC3 connector to the end of each speed controller's blue leads. Refer to Figure 3.18.
8. Connect the male EC3 connectors to the female EC3 connectors of your speed controllers and the male motor connector to the matching female on the extension and verify that they fit properly. Refer to Figure 3.19.
9. Disconnect them again for now.

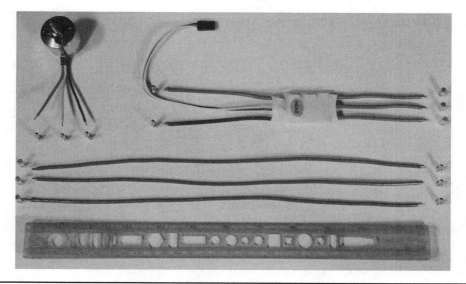

FIGURE 3.16   ESC to motor wiring.

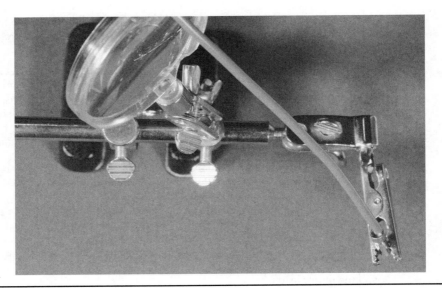

FIGURE 3.17   Soldered ESC wires.

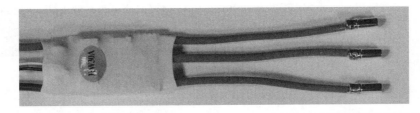

FIGURE 3.18   Female EC3s connected to an ESC.

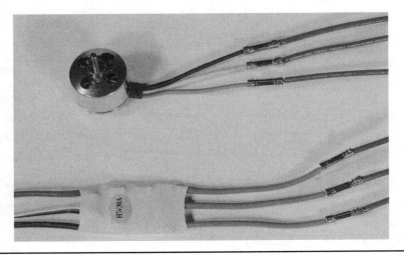

FIGURE 3.19   Wire extensions connected to an ESC and motor.

Apply Heat-Shrink Tubing to Motor and ESC Leads

Heat-shrink tubing will protect the solder joints and prevent unintended electrical contact. As shown in Figure 3.20, the tubing is shrunk over both a female EC3 wire extension connector and a male EC3 connector.

This protection keeps the leads from accidentally making contact with each other, yet allows connectors to be plugged and unplugged as needed when testing motor connections in a later step.

NOTE: *Using this technique of applying shrink tubing might leave a very small amount of exposed connector. However, there is a very small chance that any exposed areas can inadvertently short. The following approach is quick and efficient and will speed your build.*

1. Locate the 27-in (68.6-cm) length of 3/16-in (0.2-cm) (black tubing. Measure and cut it into 5/8-in (1.6-cm) pieces, which will be about 42 pieces total.
2. Slide shrink tubing over the male and female EC3 connectors soldered to the wire extension leads.
3. Plug in the opposite connector from the motor assembly. Position the tubing to cover both connectors and solder joints.
4. Carefully apply heat to all the shrink tubing on all of the connectors.
5. Leave the wire extensions plugged into the motors, but unplug the wire extensions from the ESCs for now.

Motor-Mount Assembly

In this step, you will attach the motor mounts, motors, and landing gear legs to the booms. *Do not attach the propellers to the motors yet!*

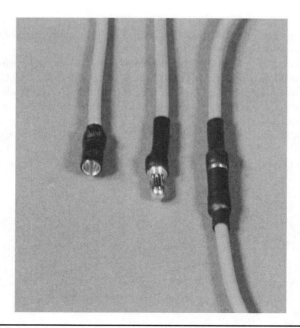

FIGURE 3.20   Heat-shrink tubing over female and male EC3 connectors.

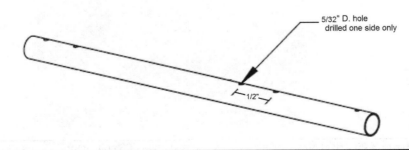

ITEM NO.	PART NUMBER	QTY.
1	Motor Assembly	1
2	1" Screw Pan head, Stainless Steel	2
3	Motor Mount top	1
4	Boom Arm Black	1
5	3/8" Screw, Pan head, Stainless Steel	4
6	Internal Tooth Lock Washer #4	4
7	Nylon Standoff 4-40 x 5/8	2
8	Motor Mount Bottom	1
9	Screw, 3mm x 6mm, 0.5 thread	4
10	Lock Nut 4-40 x 1/4	1
11	Landing Gear	1

Figure 3.21 Motor mount assembly diagram.

1. Gather the items listed in the Figure 3.21.
2. Attach each motor (item 1) to a motor-mount bottom plate (item 8). Use four 3-mm × 6-mm screws (item 9) for each motor.
3. Use two ⅜-in (1-cm) pan-head screws (item 5) and two internal tooth-lock washers (item 6) to attach two ⅝-in (1.6-cm) nylon standoffs (item 7) to each motor-mount bottom plate. Use two more ⅜-in (1-cm) pan-head screws and internal tooth-lock washers (item 6) to attach each motor-mount top plate (item 3) to the nylon standoffs (item 7).
4. *Note: I have added the following step to tidy up the boom wire installation.*
   Drill a ⁵⁄₃₂-in (0.4-cm) diameter hole that is located ½ in (1.3 cm) away (toward the motor end) from one of the two holes that are spaced 2¾ in (7 cm) apart. Refer to

5/32" D. hole
drilled one side only

1/2"

Figure 3.22 LED power lead access hole position.

Figure 3.22 for this hole position. This new hole will allow you to thread the LED power leads through the boom in lieu of being outside of the boom.

WARNING: Drill the hole on one side only. Do not drill through the boom.

5. One end of each boom (item 4) has two holes spaced about 1 in apart. Slide an assembled motor mount on this end of each boom so that the motor leads go through the boom tube. Holes in the motor-mount top and bottom plates will line up with the holes in the boom.

NOTE: Position the boom so that the extra hole drilled in step 4 is facing downward.

6. Secure each motor-mount assembly to the last hole in its boom with a 1-in (2.5 cm) pan-head screw (item 2) and a lock nut (item 10).

7. Thread a 1-in (2.5-cm) pan-head screw (item 2) through the second hole in each boom and motor-mount assembly. This screw does not use a lock nut. Instead, self-tap the screw into the shorter leg of a landing gear (item 11) aligned parallel to the boom.

Boom Accessories

Your Elev-8 Quadcopter Kit comes with two options for accessorizing the booms: checkered tape and adhesive-backed LED light strips. These accessories are optional but highly recommended. You can apply either, both, or none to the booms. Many people choose to put white LED strips and black/white tape on the front booms, and red LED strips and red/white tape on the back booms. This makes it easy to identify the front and back of the quadcopter during flight. If you wish to use both, apply the checkered tape before applying the LED light strips.

NOTE: The decal application differs from what I mentioned in Chapter 2, but it does make sense and I would concur on placing the decals and LED strips as instructed.

1. Cut each sheet of checkered tape in half lengthwise to make four pieces. Apply a piece of tape around each boom. This will make two red/white booms and two black/white booms.

2. Locate the black and red 22 AWG wires in the Elev-8 Electronics Kit and the additional wire if you have decided to add a light controller to the quadcopter.

3. For a non-light-controller installation, follow this step:
Cut each 22 AWG wire into two pieces approximately 9 in (23 cm) long, and then strip ¼ in (0.6 cm) of insulation from each end. You will have four black leads and four red leads for your LED tapes.

4. For a light-controller installation follow this step:
Cut each 22 AWG wire into two pieces approximately 14 in (35.6 cm) long, and then strip ¼ in (0.6 cm) of insulation from each end. You will have four black leads and four red leads for your LED tapes.

5. Locate the two LED tape strips. The yellowish LEDs shine white, and the clear LEDs shine red. Cut each strip in half along the solid black line. See Figure 3.23 for confirmation of where to cut. Cutting in the wrong place will destroy the strip.

6. Each LED tape section has tiny (+) and (−) contacts on one end. Solder a red 22 AWG lead to each (+) contact, and solder a black 22 AWG lead to each (−) contact. Label the ends of each wire pair to indicate which boom they are attached to, i.e., "left rear boom."

7. Thread the red and black wires through the hole that you drilled in the boom, as shown in Figure 3.22. Push the wires toward the center, away from the motors.

8. Peel the backing off of an LED tape section, and position it along the underside of a boom (over the checkered tape), with the wires pointing away from the motor.

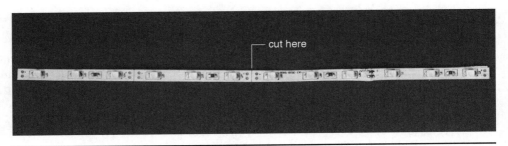

FIGURE 3.23 Cutting the LED lighting strip.

9. Measure and cut four pieces of ¾-in (1.9-cm) clear heat-shrink tubing. Each one should be 4.5 in (11.5 cm) long.
10. Slip the heat-shrink tubing over each boom to cover the LED strip and its solder joints, and apply heat to shrink it in place. Refer to Figure 3.24 to see a completed assembly mounted on a boom.

Attach Motor/Boom Assemblies to the Bottom Chassis Plate

In this step, you will attach each motor/boom assembly to the bottom chassis plate. (This kit comes with two identical quad chassis plates. You will use one in this step and the other one as a chassis top plate in a later step.) Refer to Figure 3.25 for the items needed from your Elev-8 Airframe and Hardware kits.

1. Locate the correct mounting holes in the bottom chassis plate to use for the motor/boom assemblies.
2. Position a motor/boom assembly (item 2) on the bottom chassis plate (item 3). The boom tube rests on top of the plate with the motor axle pointing upward, and the LED tape facing downward. The free end of the landing-gear leg slips underneath the plate.
3. Thread a 1¼-in (3.2-cm) long pan-head screw (item 5) through the end of the landing gear's longer leg, through the underside of the chassis plate (item 3) and through the boom tube (item 2). Secure it in place with a threaded ⅝-in (1.6-cm) nylon standoff (item 1).

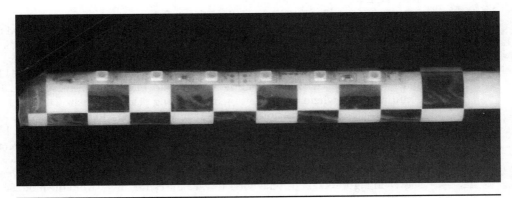

FIGURE 3.24 Completed LED strip lighting assembly.

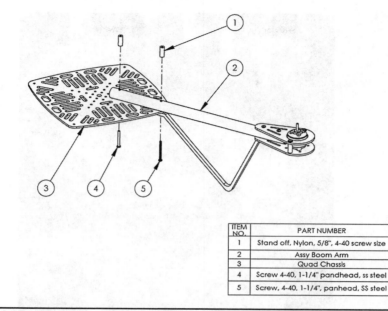

ITEM NO.	PART NUMBER	QTY.
1	Stand off, Nylon, 5/8", 4-40 screw size	2
2	Assy Boom Arm	1
3	Quad Chassis	1
4	Screw 4-40, 1-1/4" pandhead, ss steel	1
5	Screw, 4-40, 1-1/4", panhead, SS steel	1

FIGURE 3.25   Attaching motor/boom assemblies to the bottom chassis plate.

4. *NOTE: Skip this step if you are using a quad power-distribution board.*
   Thread a 1-in (2.5-cm) long pan-head screw (item 4) through the underside of the chassis plate (item 3), and then through the last hole in the boom (item 2). Secure it in place with a threaded ⅝-in (1.6-cm) nylon standoff (item 1).

5. *NOTE: Use this step for a quad power-distribution board installation.*
   Thread a 1¼-in (3.2-cm) long pan-head screw (item 5) first through one corner hole of the quad power-distribution board, then through the underside of the chassis plate (item 3), and then through the last hole in the boom (item 2). Secure it in place with a threaded ⅝-in (1.6-cm) nylon standoff (item 1). Figure 3.26 shows the quad power-distribution board installed.

6. Repeat the steps until all four motor/boom assemblies are attached to the bottom chassis plate.

Solder the Power Harness Together

NOTE: I have copied these instructions verbatim from the package instructions. Since I did not solder a power harness together, I have no photos of this process, and thus, I must refer you to the photos contained in the package instructions. As you read through the following instructions, I think you will get an appreciation of why I chose to use the quad power-distribution board. My instructions on installing the quad board follow this instruction set.

In this step, you will solder together your Elev-8 quadcopter's power harness. It will provide the connection between the battery pack and the ESCs (and LED tapes if you are using them).

1. *Find the black and red 12 AWG wires in your ELEV-8 Electronic Kit; these will be the power harness leads. Strip ¼ inch of insulation off one end of each wire.*

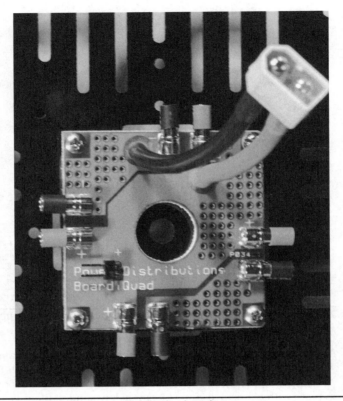

Figure 3.26   The installed quad power-distribution board.

2. *Solder all the ESC speed controllers' red leads to the single 12 AWG red lead, ends-to-end. Likewise, solder all of the ESC's speed controllers' black leads to the single 12 AWG black lead, ends-to-end, as shown in the picture at right, top.* [Author's note: See the layout in the package instructions.]

3. *Cut two 1½-inch lengths of ½-inch black tubing. Slip a piece of tubing past each of the two solder joints you just made on the power harness so that they sit closest to the ESCs. Do not shrink them yet, just keep them out of the way of the solder joints.*

    Position the power harness inside the ELEV-8 chassis bottom, but do not secure it in place yet. The recommended layout is shown on the following page. [Author's note: The layout is in the package instructions.]

4. *If you are using LED tapes, bundle together all of their red leads. Align these thinner wires alongside the red 12 AWG wire in the opposite direction from the ESCs' wires, and solder them into place, as shown at right, middle.* [Author's note: See the layout in the package instructions.] *This will make a neater package for the heat shrink tubing.*

5. *Likewise, solder the LED tapes' black leads to the solder joint where all the other black leads meet on the power harness.*

6. *Slide the heat shrink tubing back up and over each solder joint. Apply heat with a heat gun or hair dryer to shrink the tubing into place, as shown at right, bottom.* [Author's note: See the layout in the package instructions.]

7. *Decide how long to trim the 12 AWG power harness leads. If you use the layout shown below* [Author's note: The layout is in the package instructions] *and you will be*

strapping your batteries to the top of the chassis, you can trim the power harness leads to about 4 inches. If you are going to use a custom layout to accommodate extra electronics, decide how long to make the power harness leads. Trim the 12 AWG wires to the desired length, and then strip about ⅛ inch of insulation from the end of each one.

8. *Locate the packet of gold bullet connectors and plastic housings in the ELEV-8 Electronics Kit. You will need two bullet connectors and one plastic housing.*

9. *Solder a bullet connector onto the end of each 12 AWG lead. Insert the exposed tip of the wire into the cup end of the bullet connector, and fill the cup with solder.*

10. *Insert the bullet connectors into the flattened end of the blue housing. The red lead goes into the "D" shaped side and the black lead goes into the "O" shaped side. It will take some force for the bullet connectors to click into place.*

    PRO TIP: *We recommend using a flat-head screwdriver to hold the bullet connector in the housing and then use a hammer and tap the connector into place.*

11. *Reposition the power harness inside the ELEV-8 chassis bottom. Secure the ESCs in place with zip ties.*

12. *Connect the battery pack to the power harness. The LED tapes, if you are using them, should now come on.*

Installing the Quad Power-Distribution Board

This instruction set assumes you have chosen to purchase a quad power-distribution board and are ready to install it. The installation is quite simple in comparison to building your own power-distribution harness.

1. The power-distribution board should already be mounted on the bottom chassis plate, as shown in Figure 3.26.
2. Tie-wrap all the ESCs down, and thread both the red and black ESC power leads through the bottom chassis plate, as shown in Figure 3.27.

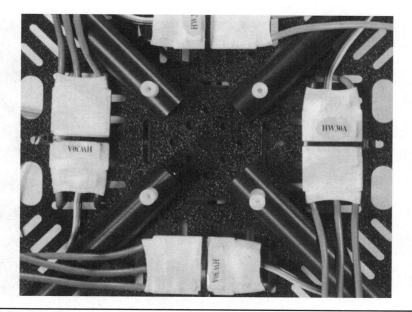

FIGURE 3.27   The ESCs mounted and the power leads threaded.

3. Solder a male EC3 connector to the end of each speed controller's red and black leads.
4. Place a ⅝-in (1.6-cm) piece of shrink tubing so that it covers the solder connection but does not overlap onto the male connector.
5. *Note:* I used red shrink tubing for the red leads and black shrink tubing for the black leads. It is not very important to do this, yet it does lend a nice touch to the installation.
6. Connect all the ESC power leads to the power-distribution board, ensuring that the red leads are inserted into the red EC3 board connectors and the black leads into the black EC3 board connectors. See Figure 3.28 showing all the ESC power leads connected to the quad power-distribution board.
7. Leave all the LED wiring unconnected for now. I will show you how to wire the LED strips in the next chapter.
8. Figure 3.29 shows all the ESCs connected to their respective motor wire extensions. Additionally, I added extra ty-wraps to help minimize the wiring confusion. I also labeled all the BEC cables and all the LED-strip wires, which will greatly help when they are connected to the flight controller and LED controller, respectively.

Configure Your Transmitter

I used a Spektrum DX-8 R/C transmitter shown in Figure 3.30 for the configuration. Other model R/C transmitters will be similarly configured because most conform to the same R/C manufacturer standards. For best results, follow the transmitter setting recommendations in Table 3.1 to configure your transmitter. Refer to Figure 3.31 to see how your transmitter's 2-axis joystick controls will translate into Elev-8 quadcopter motion with these settings.

Note: I have discussed this "translation" in depth in the section on "Flight Controls" in Chapter 2. I just included the package instructions to maintain consistency.

Figure 3.28 ESC leads connected to the quad power-distribution board.

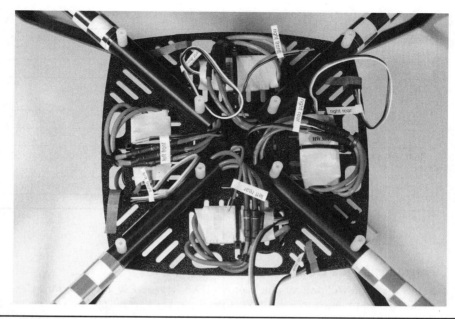

Figure 3.29   All the ESCs connected to the motor wire extensions.

Figure 3.30   Spectrum DX-8 R/C transmitter.

Box model type	ACRO (Plane Mode)
End point adjustment	Set to 50% initially. (If the Elev-8 still seems too reactive, reduce to 30% until you get used to flying it.)
Dual-rates (D/R)	100%
Channel reverse	Normal: Hi Tech Spektrum, JR Brands Reversed: Futaba brand
Trims	Centered
Sub-trims	Centered
Gain adjust	Set gain on 5th channel. Start with 25%; add or subtract based on flight stability.
Exponential	After gaining experience, add up to 30% into aileron and elevator.

TABLE 3.1 Transmitter Setting Recommendations

Programming the Electronic Speed Controllers

In this step, you will program the motor's ESCs with an ESC Programming Card. The ESCs should not be plugged into the motors yet. *If the ESCs are plugged into the motors, disconnect them now.*

NOTE: *The ESC Programming Card was added to the kits in April 2013. They are also available separately from http://www.parallax.com; search for "85000."*

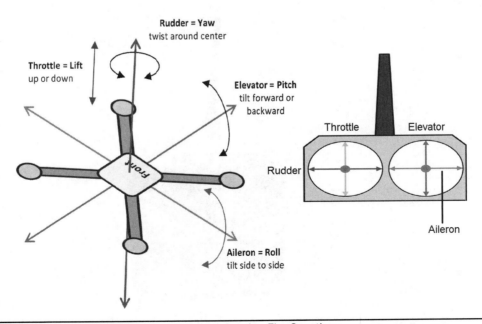

FIGURE 3.31 Transmitter 2-axis controls translated to Elev-8 motion.

1. Connect an ESC to the ESC programming card's BEC port. Be sure to line up the black wire with (−), the red wire with (+), and the white wire with (Signal).
   WARNING: Do not connect to the Programming Card's BEC port and Batt port at the same time; this would damage the card.
2. Connect your charged LiPo battery to the power harness.
   NOTE: I changed the sequence from the original order shown in the Parallax instruction sheets. The programming card is not recognized if you connect the LiPo battery first and then connect the card to the ESC. This is probably due to the default ESC initialization sequence.
3. Set the ESC card to the configuration shown in Figure 3.32 and Table 3.2, and then push OK to program the ESC. Repeat with each ESC, using the same settings. Be sure to cycle power between each programming cycle.

Connect the Motors and Synchronize the ESCs

After programming the ESCs, it is time to connect each one to its motor and test it to make sure it is turning in the proper direction. For this step, your receiver will temporarily connect directly to (and receive power through) each ESC.

WARNING: Do not connect a battery or other power source and an ESC to your receiver at the same time. If you do, you will permanently and catastrophically damage both the ESC and receiver.

STOP: You should not have propeller blades on your motors yet! If you do, remove them now!

1. If you have not done so already, bind your transmitter to your receiver as per your R/C controller's instruction manual.

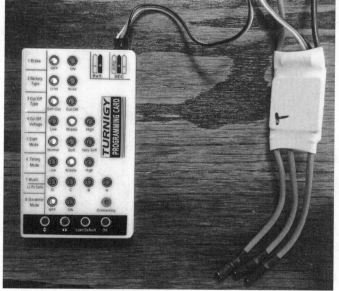

FIGURE 3.32 ESC programming.

Brake	Off
Battery type	Li-xx
Cut-off type	Soft-cut*
Cut-off voltage	Middle
Start mode	Normal
Timing mode	Middle
Music/Li-Po cells	(none)
Governor mode	Off

*Soft-cut (also called Reduce Power) lets you know when the quadcopter's batteries are running low. If you set this to Cut-off/Shut Down, your quadcopter will simply fall out of the sky when it reaches a certain battery level.

TABLE 3.2 ESC Programming Parameters

2. Identify which edge of the chassis will be the front of your Elev-8 quadcopter. If you have used the checkered stickers and/or the LED tapes, the front edge would be between the two black-checkered, white-LED booms.
3. Put a piece of tape on the output shaft of each motor so that you can easily tell the direction of rotation.
4. Connect an ESC's 3-pin socket to the throttle port on your receiver.
5. Gently apply the throttle to see in which direction the motor turns. Refer to Figure 3.33 to identify the direction in which each motor needs to turn.
6. If the motor is not turning in the proper direction, disconnect any two of its leads, reverse them, and reconnect.
7. Label the ESC with its motor position number, both on its case and on its 3-pin socket.
8. Repeat with each ESC until all motors are turning in the correct direction and each ESC case and lead are numbered.

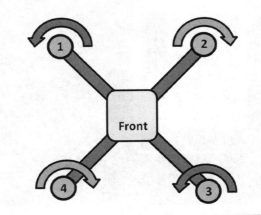

FIGURE 3.33 Motor rotation.

9. When you are sure your motor connections are all correct, apply heat to finish shrinking the tubing over the motor/ESC connector joints.
10. To synchronize the ESCs, power on the Elev-8 quadcopter. Turn on your transmitter, then set the throttle to max position. After the standard startup sequence, two separate beeps will indicate that the max throttle position has been set and stored. Lower the throttle to min position. You will hear three beeps, which indicate that min throttle position has been set and stored.

Chassis Top-Plate and Control-Board Assemblies

NOTE: I would recommend reading Chapters 7 and 8 now if you are considering installing a video system for use with your quadcopter. Mounting the video camera frame assembly would be easier at this point with the chassis top plate not yet installed. You certainly could continue with the assembly without a video camera frame installation but eventually you will have to disassemble your quadcopter back to this point if you later decide to install it.

In this step, you will prepare and attach the chassis top plate. Then, you prepare and attach the control board to its mount plate.

NOTE: The control board mount plate has slots around all four edges.

1. Gather the items shown in Figures 3.34 and 3.35.
2. Pull all the ESCs' 3-pin leads together towards the front of the chassis.
3. Refer to Figure 3.33 to locate the correct holes in the chassis top plate (item 2) in order to attach the four 1¼-in (3.2-cm) nylon standoffs (item 1).

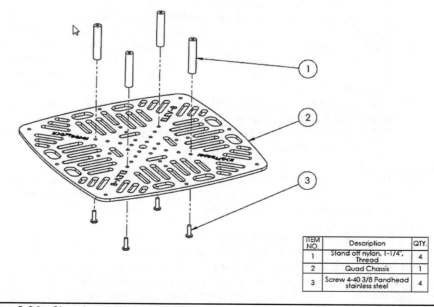

ITEM NO.	Description	QTY.
1	Stand off nylon, 1-1/4", Thread	4
2	Quad Chassis	1
3	Screw 4-40 3/8 Pandhead stainless steel	4

FIGURE 3.34 Chassis top-plate assembly.

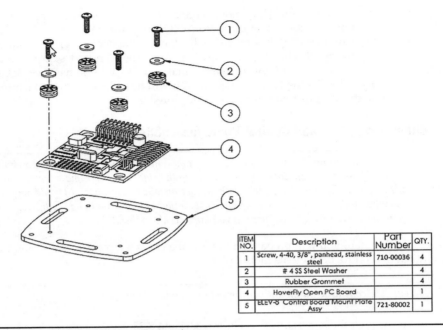

ITEM NO.	Description	Part Number	QTY.
1	Screw, 4-40, 3/8", panhead, stainless steel	710-00036	4
2	# 4 SS Steel Washer		4
3	Rubber Grommet		4
4	HoverFly Open PC Board		1
5	ELEV-8 Control Board Mount Plate Assy	721-80002	1

Figure 3.35   HoverflyOPEN board mounting.

4. Attach each nylon standoff (item 1) to the top of the chassis top plate (item 2) with a ⅜-in (1-cm) pan-head screw (item 3).
5. Attach the chassis top plate to the standoffs on top of the booms using ¼-in (0.6-cm) black pan-head screws. There will be two screws required for each boom.
6. Refer to Figure 3.35. Rubber grommets are included with the control board. Insert a rubber grommet (item 3) into the large mounting hole on each corner of the control board (item 4). These grommets reduce vibrations transferred to the control board during flight.
7. Insert each ⅜-in (1-cm) pan-head screw (item 1) through a washer (item 2), then through an installed grommet, and finally, into the control-board mount plate (item 5). The screws are self-tapping into the mount plate, so only gently hand-tighten to avoid stripping the hole.

Note: *You may want to use ½-in (1.3-cm) 4-40 nylon screws and nuts in lieu of the steel pan-head screws described in step 6 above. Figure 3.36 shows the top of the control board mounted with the nylon screws. Notice that I did use the steel washers. The underside of the chassis board mount plate is shown in Figure 3.37.*

You can clearly see the nylon 4-40 nuts attached to the nylon screws. There are two advantages to changing from the steel pan head screws to the nylon screws. First, because nylon is much less stiff than steel, the nylon screws will transmit much less vibration to the control board. The second advantage is that using nylon nuts is a more secure way of fastening the control board to the mount plate compared with using ordinary machine screws to self-tap into the mount board.

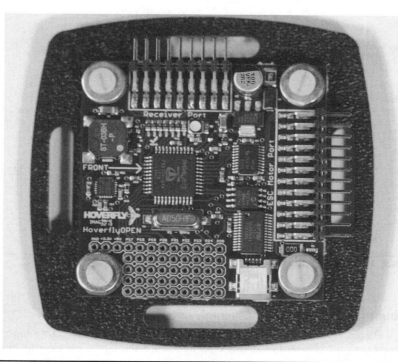

Figure 3.36   Control board mounted with nylon screws.

Figure 3.37   Underside of the chassis-board mount plate.

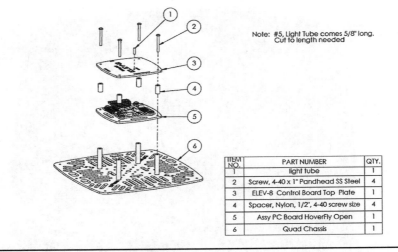

Note: #5, Light Tube comes 5/8" long.
Cut to length needed

ITEM NO.	PART NUMBER	QTY.
1	light tube	1
2	Screw, 4-40 x 1" Pandhead SS Steel	4
3	ELEV-8 Control Board Top Plate	1
4	Spacer, Nylon, 1/2", 4-40 screw size	4
5	Assy PC Board HoverFly Open	1
6	Quad Chassis	1

FIGURE 3.38   Control-board mount assembly.

Mount the Control-Board Assembly onto the Chassis

In this step, you will enclose the control board within its protective top plate, and then mount the control-board assembly onto the completed Elev-8 quadcopter chassis.

1. Gather the items listed in Figure 3.38.
2. Find the arrow on the control-board silkscreen, as shown in Figure 3.39. This arrow points to the front of the control board, which must be facing the same direction as the front of the Elev-8 chassis.

FIGURE 3.39   Mounted HoverflyOPEN board (with steel screws).

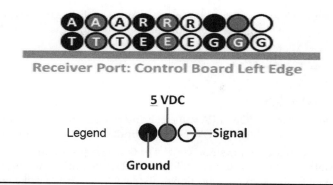

Receiver Port: Control Board Left Edge

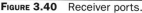

FIGURE 3.40   Receiver ports.

3. Set the control-board assembly over the standoffs in the chassis top plate. Make sure the front of the control board is aligned with the front of the chassis.

4. Align the control-board top plate over the control board. The small hole near the center of the control-board top plate is for a light pipe. Make sure this hole is directly above the LED on the control board. This will make light from the LED visible when the top plate is in place.

5. Thread each 1-in (2.5-cm) pan-head screw (item 2) through a corner hole in the control-board top plate (item 3), then through a ½-in (1.3-cm) nylon spacer (item 4), then through a corner hole in the control-board assembly (item 5), and finally into a standoff on top of the chassis (item 6). Gently tighten.

6. Insert the light pipe (item 1) into its hole in the control-board top plate (item 3), until it touches the LED underneath. Trim the light pipe to length.

7. Slip your battery between the control-board mount plate and the chassis top plate, and secure it in place with the nylon straps.
   NOTE: If you are using the quad power-distribution board, you might find it more convenient to mount the battery under the bottom chassis plate but over the quad board. You will need to use ⅜-in (1-cm) spacers between the battery and the bottom chassis plate to avoid squashing the quad board.

8. Mount your receiver to the chassis with zip-ties, referring to its documentation for best placement recommendations.

Control-Board Connections

In this step, you will connect your ESCs and receiver to your control board. The receiver connects to the receiver port's 2 × 9 male header on the left edge of the control board, as shown in Figure 3.40. The electronic speed controllers connect to the ESC port's 2 × 12 male header on the front edge of the control board, as shown in Figure 3.41.

ESC Port: Control Board Front Edge

FIGURE 3.41   ESC ports.

1. Connect the receiver to the receiver port, with the five signal connections listed below. Use the 3-wire extension cables included in the Elev-8 Electronics Kit.

    A = Aileron

    T = Throttle

    R = Rudder

    E = Elevator

    G = Gear   (ON: EPA value is Primary Gain, Altitude Hold is off.)
                  (OFF: EPA value is Altitude Hold Gain, Altitude Hold is on.)

2. Connect each motor's ESC controller to the corresponding pins on the ESC port.
3. Match the motor numbers in Figure 3.32 to the port numbers in Figure 3.41.
4. Double-check your connections—it's easy to make a mistake here.

Mounting the Propeller Blades

CAUTION: Mount the propeller blades only when you are ready to fly.

I would highly recommend that you balance all of the propellers before mounting them. Balancing the propellers will greatly reduce the vibrations that result from slight imbalances that are likely to be present in the plastic propellers supplied with the kit. Even very slight imbalances create considerable vibration because of the very high propeller rotations. Reducing vibrations will also improve the performance of the in-flight video system described in Chapter 8. I used a magnetic propeller balancer from Top Flite, which is shown in Figure 3.42.

FIGURE 3.42   Magnetic propeller balancer.

The propeller is mounted on a steel rod that is effectively suspended between two very strong magnets. One end of the rod contacts a magnet, while there is a ⅟₃₂-in (0.1-cm) gap between the other rod end and the other magnet. You can barely see this gap on the left side of the rod in the figure. This method of suspending the rod is practically free of friction and allows the propeller to swing freely so that the heavier side always swings down. I very (and I mean *very*) lightly sanded the back side off the propeller's heavy side until the propeller no longer had a heavy side. I took the propeller off the balancer, sanded a little bit, and then put the balancer back in until it no longer rotated downward. This is a tedious process but well worth the effort.

There are two different types of slow-flyer propeller blades in the Elev-8 Electronics Kit: counterclockwise, (CCW, marked 1045) and clockwise (CW, marked 1045R). The correct type of blade must be used on each motor for the Elev-8 to fly. See Figure 3.43 for label location; the blades are rounded side up.

1. Disconnect the battery from the power harness.
2. Refer to Figure 3.44 for the correct placement of each blade.
3. Referring to Figure 3.45, connect each blade to its motor . The blade (item 2) should be mounted rounded-side-up, seated on a cone lock (item 3) over a collet (item 4).
4. Finger-tighten the propeller nut (item 1), and then use an Allen wrench to tighten it ¼ turn more.

Mounting the Battery

The battery you select to be used with your Elev-8 must be mounted securely to the quadcopter. Two Velcro™ straps shown in Figure 3.46 are provided with the kit with the intention that they be used to fasten the battery to the bottom side of the bottom chassis plate.

The straps should be threaded through any convenient chassis to secure the battery to the plate. However, existing power cables may interfere with a tight strapping especially if the power-distribution board is already mounted. My solution is to use a battery mount plate that is raised above the bottom chassis plate, using ½-in (1.3-cm) nylon spacers. Figure 3.47 is a construction diagram for the Lexan™ battery mount plate.

The battery mount plate is secured to the bottom chassis plate using two flat 6-32 1-in (2.5-cm) machine screws with matching nuts. Countersinking the screw holes in the Lexan™

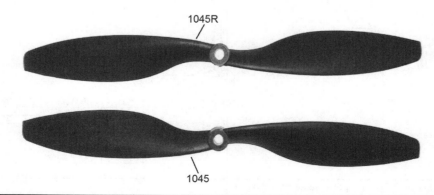

Figure 3.43 Propeller type identification.

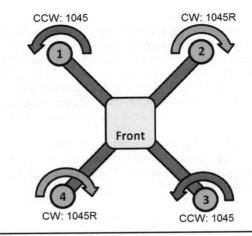

Figure 3.44   Propeller placement.

plate ensures that the battery will lie flat on the mount plate. Figure 3.48 shows the installation of a completed battery mount plate without the battery so that you can clearly see the mounts I used and their placement on the bottom chassis plate. Now you can strap the battery to the mount plate very easily without any interference from other components or wiring.

The following paragraph is the end of the package instructions.

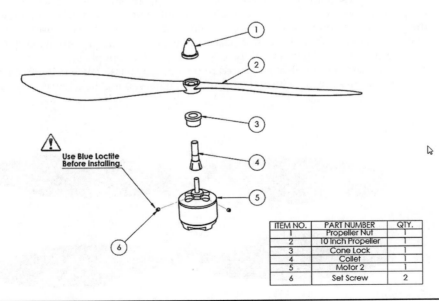

Figure 3.45   Propeller assembly diagram.

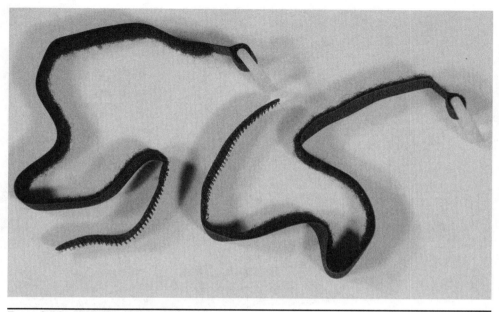

FIGURE 3.46   Velcro™ mounting straps.

Congratulations! Your assembled Elev-8 Quadcopter is ready to fly. For a "First Flight" video and troubleshooting tips, see the ELEV-8 Quadcopter product page; go to www.parallax.com and search for "80000."

However, I have a few more comments on the build before the chapter finishes.

A Few More Comments

You should now be in possession of a fully functioning Elev-8 quadcopter. However, I would urge you not to fly just yet because there are still some important modifications that should

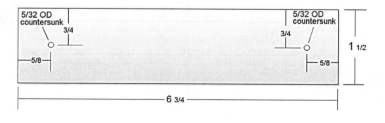

material: 3/32 Lexan

FIGURE 3.47   Lexan™ battery mount plate construction diagram.

FIGURE 3.48   Completed battery mount plate without battery.

be done. It would also help you to read the material in the following chapters to expand your knowledge of operating and modifying the quadcopter system.

I also feel strongly that you should incorporate the "kill switch" functionality discussed in Chapter 10, which will enable you to instantly disable the quadcopter in the unlikely event it becomes uncontrollable and is heading for a person or property. It is much better to drop the quadcopter out of the sky then to seriously harm a person or cause significant property damage.

CHAPTER 4

Programming the Parallax Propeller Chip

Introduction

The Parallax Propeller chip that was introduced in Chapter 1 will be discussed in this chapter from hardware and software perspectives. I will also refer to the chip as the *Prop* from now on for brevity's sake.

The Prop was developed by Parallax engineers solely for an extended project that started in the late 1990s and ran to its market introduction in 2006. It is still the only available multicore processor designed specifically for the technical/hobbyist market. Intel and AMD multicore processors are distinctly different technologies designed for PC and server applications. The Prop also has a much lower clock speed than Intel processors, which is not an issue because it serves a very different purpose than the Intel/AMD multicores do. Props are also very inexpensive, typically less than $10 USD, making them ideal for experimentation and prototyping. First, I will describe the Prop's unique architecture before going on to cover the software and programming details.

Prop Architecture

In Figure 4.1 is the Prop P8X32X block diagram showing the eight separate cores along with other key elements.

One of the first things that you might have noticed is that the cores are called cogs in the Prop diagram. Each cog is an independent processing element with its own 2-kB memory. The Prop also has 32-bit data and address busses, which means the "word" size is also 32 bits. There are 32 *general-purpose input/output* (GPIO) pins that each cog can access. These are shared GPIO pins; however, some or all may be dedicated to a specific cog or group of cogs if you want. A GPIO pin so dedicated is also referred to as being mutually exclusive indicating that the pin is being controlled by a designated cog.

All the cogs communicate with a hub that contains a common or shared 64-kB memory that is divided as 32 kB of RAM and 32 kB of ROM. The hub also contains configuration, power management, reset, and clock circuits.

The clock circuits are quite flexible and may use either an internal *resistor-capacitor* (RC) oscillator or a crystal-controlled oscillator, which in turn requires an external crystal. Most Prop boards that I have used operate with a crystal oscillator because it provides higher

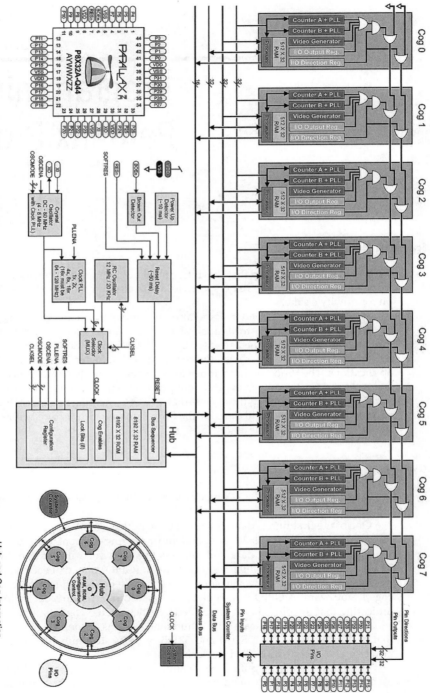

FIGURE 4.1 Prop block diagram.

speed, better frequency stability, and better precision of the clock signal than the built-in internal RC oscillator does. In addition, the applications discussed later in this chapter require precise timing, which necessitates the use of the crystal-controlled oscillator.

The Prop chip also has a *phase-locked loop* (PLL) circuit that is used only in conjunction with the crystal oscillator. The PLL circuit can multiply the external crystal resonant frequency in steps of $1\times, 2\times, 4\times, 8\times$ or $16\times$. It is very common to have a Prop board equipped with a 5-MHz crystal used with the $16\times$ PLL multiplier to create an 80-MHz Prop clock frequency. This does not quite match the 2.4- to 3.0-GHz range of Intel processors, but it does not need to in order to meet the needs of the Prop embedded applications.

The Prop may also be thought of as a microcontroller because of the GPIO pins and because each cog contains two versatile counters that may be configured in different ways to suit an application. Every cog also has its own video generator circuit that both improves display performance and increases program display functionality.

The hub and cogs interface in a round-robin fashion, as may be seen in Figure 4.1. Each cog has a time slice in which to access the common 32-kB RAM memory. The hub-memory access time is set to be 50% of the system-clock speed, which means a 40-MHz access rate for an 80-MHz clock rate. Cogs are also rated at a nominal 20 *million instructions per second* (MIPS) when operating at the 80-MHz clock speed. The MIPS rating is a result of an average Prop instruction taking four clock cycles to complete. The 40-MHz round-robin hub access speed means that data transfers between cogs and the shared hub memory are not constrained or limited.

The 32-kB hub memory ROM is used to store common data tables as well as the character generator set that each cog video generator can utilize to display programmed video output. Sharing a common character data table frees up the individual cog memory for its own program. 8 kB of hub ROM is also reserved for the Spin interpreter that converts source-level Spin code into executable native instructions or tokens. A space-optimized copy of the Spin interpreter is put into an individual cog's memory, thus enabling the cog to perform real-time processing of the Spin tokens designated to run in that cog. The Spin language is discussed in detail below in the programming section.

NOTE: You should realize that no code is ever executed from hub memory; all code that is run in a Prop chip is executed by the cogs.

The Prop is manufactured in several physical packages including a 44-pin *low profile quad flat package* (LQFP) surface-mount configuration that has 0.8-mm pin spacing. There is also a 44-pin *quad flat no-leads* (QFN) carrier format, and finally, a 40 *dual inline package* (DIP) suitable for solderless breadboard prototyping. I used the Parallax Propeller *Board of Education* (BOE) shown in Figure 4.2 to develop and test the software used for the auxiliary functions for the Elev-8 quadcopter.

The BOE has an LQFP-mounted Prop chip that is connected to many different peripherals, which allows for rapid and convenient Prop software development. This is certainly true for the servo-motor test application discussed in Chapter 5. There are a series of servo ports located at the top right of Figure 4.3, which shows an annotated BOE photo.

There is also an external 64-kB EEPROM installed on the BOE that enables you to load and store programs in a nontransient state, which means that the programs remains in the EEPROM even if the power is shut off. Programs stored in RAM are considered transient because they disappear once the power is removed. The BOE board supports the Prop firmware feature whereby a program stored in EEPROM will automatically be copied into the Prop RAM and then executed once the power is applied. Please be aware that there is a

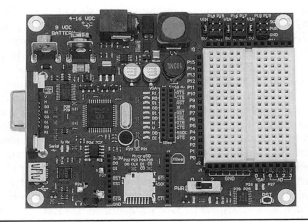

FIGURE 4.2 Parallax Propeller Board of Education (BOE).

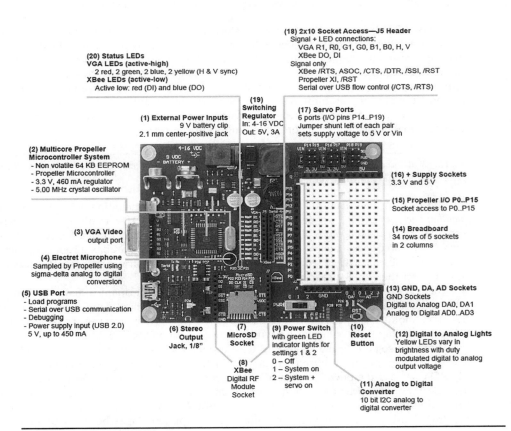

(20) Status LEDs
VGA LEDs (active-high)
  2 red, 2 green, 2 blue, 2 yellow (H & V sync)
XBee LEDs (active-low)
  Active low: red (DI) and blue (DO)

(18) 2x10 Socket Access—J5 Header
Signal + LED connections:
  VGA R1, R0, G1, G0, B1, B0, H, V
  XBee DO, DI
Signal only
  XBee /RTS, ASOC, /CTS, /DTR, /SSI, /RST
  Propeller XI, /RST
  Serial over USB flow control (/CTS, /RTS)

(19) Switching Regulator
In: 4–16 VDC
Out: 5V, 3A

(1) External Power Inputs
  9 V battery clip
  2.1 mm center-positive jack

(17) Servo Ports
6 ports (I/O pins P14..P19)
Jumper shunt left of each pair
sets supply voltage to 5 V or Vin

(2) Multicore Propeller Microcontroller System
- Non volatile 64 KB EEPROM
- Propeller Microcontroller
- 3.3 V, 460 mA regulator
- 5.00 MHz crystal oscillator

(16) + Supply Sockets
3.3 V and 5 V

(15) Propeller I/O P0..P15
Socket access to P0..P15

(3) VGA Video
output port

(14) Breadboard
34 rows of 5 sockets
in 2 columns

(4) Electret Microphone
Sampled by Propeller using
sigma-delta analog to digital
conversion

(13) GND, DA, AD Sockets
GND Sockets
Digital to Analog DA0, DA1
Analog to Digital AD0..AD3

(5) USB Port
- Load programs
- Serial over USB communication
- Debugging
- Power supply input (USB 2.0)
  5 V, up to 450 mA

(6) Stereo Output Jack, 1/8"

(7) MicroSD Socket

(9) Power Switch
with green LED
indicator lights for
settings 1 & 2
0 – Off
1 – System on
2 – System +
  servo on

(10) Reset Button

(12) Digital to Analog Lights
Yellow LEDs vary in
brightness with duty
modulated digital to analog
output voltage

(8) XBee
Digital RF
Module
Socket

(11) Analog to Digital Converter
10 bit I2C analog to
digital converter

FIGURE 4.3 Annotated BOE.

short delay once power is applied to the BOE, since it takes 1.5 seconds to copy the 32-kB ROM image into the 32-kB RAM memory.

Just in case you were wondering, user programs cannot be loaded into the Prop chip's internal 32-kB ROM memory. Doing so would overwrite and destroy the Prop's firmware, essentially eliminating the Spin interpreter and critical data storage tables—in effect, "bricking" your Prop chip.

Prop Software

Until recently, the Prop has been programmed by using the Spin language and the *Propeller Assembly Language* (PASM). Now, the C Language has been added to the available programming tools that support the Prop. I will focus initially on Spin, since it is the primary language to use when programming the Prop. I will discuss PASM and C in Chapter 5.

Spin Language

The Spin Language is one that I refer to as an *object-oriented* (OO) hybrid language, meaning that it uses many of the features of a full blown OO language, such as Java or C++, while directly dealing with the constraints and limitations inherent in a real-time programming environment. For instance, the Spin language does not support some basic OO functionality, such as creating and destroying dynamic objects. Spin's underlying reliance on the Prop chip's functionality precludes supporting this basic OO feature. This does not mean that Spin is not a highly useful OO tool; it just means that the normal programming approach must be altered to reflect the Prop's architecture. The Spin language designers recognized that not all OO programming paradigms could be realistically implemented in a real-time, parallel-processing environment. The Spin compiler is an extremely well-designed software development tool that enables you to create true parallel-processing programs that also efficiently execute in a real-time environment. There are other parallel-development tools available, but none that I know of that are so user friendly and make it so easy to develop practical program solutions.

It would be naïve of me and a disservice to you to attempt to thoroughly cover OO programming fundamentals in this chapter. I would urge you to learn some basic imperative programming concepts, and at least, go through an introduction to OO programming principles. It will be difficult for you to fully comprehend how the example Prop programs function without some programming background. If you lack a good foundation, you will find it difficult to modify the examples to suit your own needs. The authors of the *Parallax Propeller User's Manual* also make the same assumption that the reader should have some OO background. With this disclaimer stated, I recommend that everyone download and read the latest *Propeller User's Manual*. Readers who have some OO development experience will still need to read the important information within the manual. Writing Prop software is really quite different from developing normal OO software because you have to consider the presence of parallel-processing cogs and a real-time operating environment. Given these issues, I will include some detailed explanations in the following discussions in order to clarify the Spin programming statements.

Propeller Spin Tool

The *Propeller Spin Tool* (PST) is a free tool provided by Parallax, which allows you to create, load, and execute software on any Prop-based development board or functioning prototype Prop circuit. As mentioned earlier in the chapter, I used the Prop BOE for software

FIGURE 4.4   Information Dialog box showing the active USB comm port.

development. The BOE is set up to communicate with a Win 7 laptop running PST version 1.3.2. There should be a USB comm port driver automatically loaded to allow your PC to detect and communicate with whatever Prop development board you choose to use. Clicking on the *Run* drop-down menu will reveal an *Identify Hardware* choice. Clicking on this will pop up an *Information Dialog* box, as shown in Figure 4.4.

I decided to go through some of the tutorials that are carefully detailed in the *Propeller User's Manual*. I will start specifically with the Blinker1 program. I did have to slightly modify the original program that is shown in the manual in order to accommodate the available GPIO pins on the BOE. Figure 4.5 is a screenshot of the modified Blinker1 program.

This figure shows the PST development screen that Parallax calls its *Integrated Explorer*. The Explorer's central portion is an editor area where the Spin code is input and modified. There are some two pane areas on the left side of the Explorer that show the current directory along with the files contained in that directory. I deliberately selected the _Demos directory, which is part of the Propeller Library. You can easily see a variety of Spin programs that are available in the file-listing pane. Incidentally, the Propeller Library is part of the PST download.

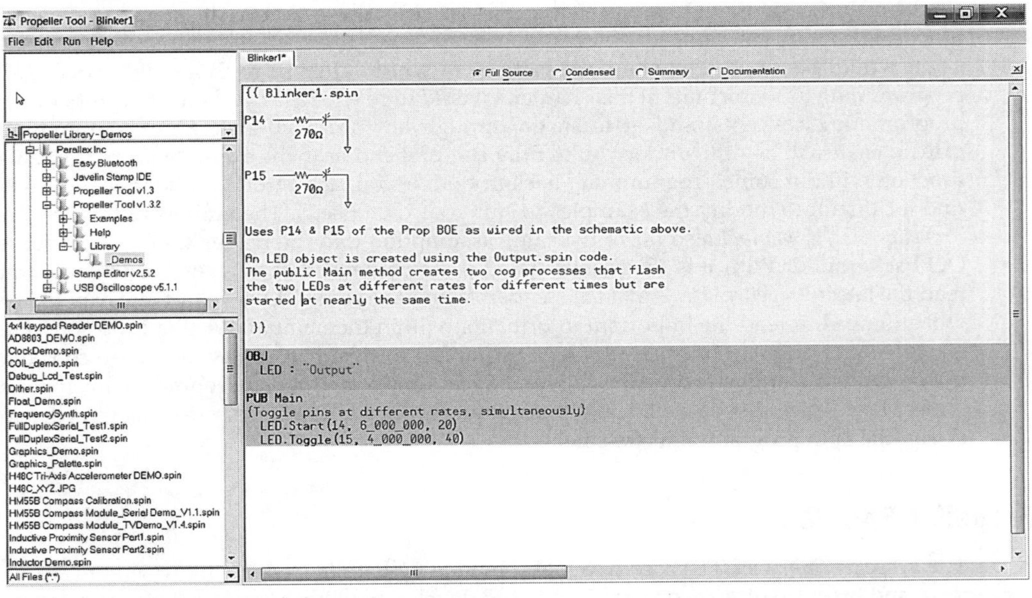

FIGURE 4.5   Blinker1 Spin program.

Comment Symbology	Purpose
{ ...comments... }	Multiline code comments (one curly brace pair)
{{ ...comments... }}	Multiline documentation comments (two curly brace pairs)
' ...comments...	Single-line code comment (one apostrophe)
'' ...comments...	Multiline documentation comments (two apostrophes)

TABLE 4.1   PST Comment and Documentation Symbols

The Blinker1 program shown in the editor portion of Figure 4.5 contains three sections:

1. Comments/Documentation
2. OBJ
3. PUB

The first section, Comments/Documentation, contains the program name along with any comments and/or explanations of what the program is supposed to do and how it works. I strongly believe that comments are a very important part of any program—even those that you write only for your own use. It is easy to imagine that you could create an uncommented program, put it aside, return to it six months later, and then not have any idea as to the function of the program. I really feel so strongly about this subject that I will not even grade any of my students' programs unless they contain comments.

The PST provides several means for adding comments to the source code. These are shown in Table 4.1. The PST uses document comment symbols to determine if a comment or a set of comments are to be shown in the Documentation View. Figure 4.6 is the Documentation View of the Blinker1 program.

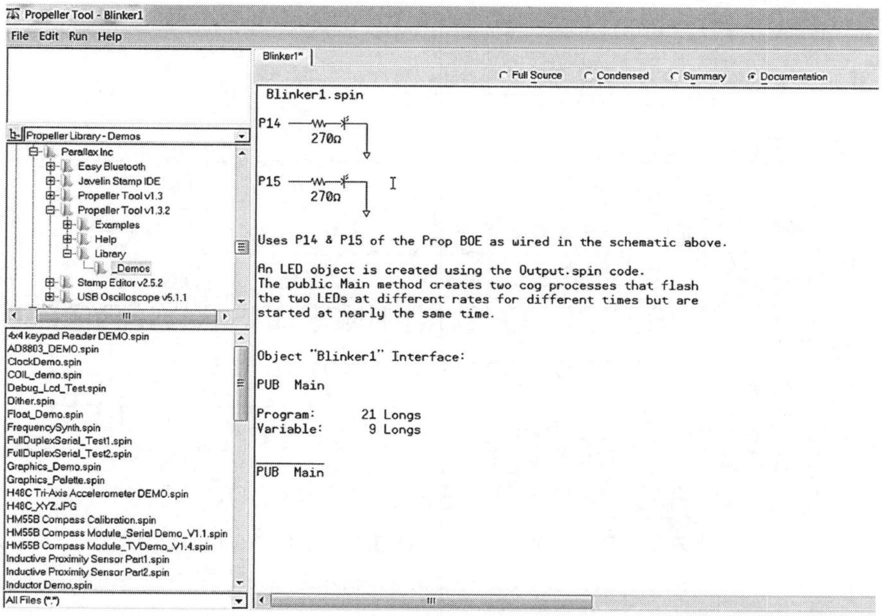

FIGURE 4.6   Blinker1 documentation view.

There are four separate views that may be selected in the PST by clicking on any of the *radio* buttons located at the top of the Editor pane. You should notice that the Full Source button is selected in Figure 4.5, while the Documentation button is selected in Figure 4.6. Condensed and Summary are the two other views that are available, but they generate very limited information, as compared to the Full Source or Documentation views for this small program. I would imagine that Condensed or Summary views might be useful for very large projects containing many objects with much more corresponding source code.

The PST also contains a comprehensive character set that includes the schematic fragments that I used to create the LED schematics shown in the top comment section of the source code. Figure 4.7 is a screenshot of the Character Chart with the Horizontal Line symbol selected.

To insert schematic characters, first position the Editor prompt at the location where you desire them to be inserted and then click on *Help* followed by clicking on *Show Character Chart*. Next, by clicking on a character in the chart, you will place that character at the location of the Editor's prompt that you had previously set. It just takes a little practice to develop your skill at creating small-scale schematics with the PST Character Chart.

The OBJ section in the Blinker1 program is where additional objects that are needed to support the program are declared. Blinker1 is also referred to as a *Top Object*, as it contains the start of program execution and also has any additional objects referenced within its OBJ section. Spin programs can have only one Top Object but may contain zero to many supporting objects. The OBJ section is how Spin establishes the object hierarchy. Other non-top objects may have their own OBJ sections to further reference more objects. The PST keeps track of this hierarchy and will issue an error if an object is missing. Referring back to Blinker1's OBJ section, you can see that it has one line in the section:

```
LED : "Output"
```

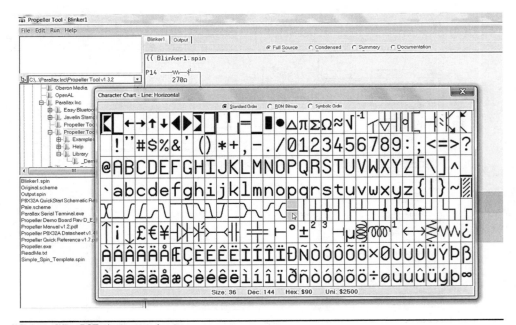

FIGURE 4.7   PST character chart.

The referenced object file name is formally Output.spin; however, the *.spin* suffix is assumed and does not need to be in the instruction, nor will it be used any further in my discussion. LED is the internal or logical reference by which the Blinker1 program will refer to the Output program. You might think of LED as a nickname in much the same way as you normally refer to friends and family by their familiar names instead of their formal names.

The last section in the Blinker1 Editor area is named PUB (which is short for PUBLIC). There is also another identifier next to PUB, which in this case is called Main. Main is the name of a method, which is something that performs an action. It can be thought of as a function or subroutine if that fits your experience. The method named Main, being the first one in the program listing, is where the Spin program will start its execution. Note that by convention only one Main method should be in a Spin program and that it is ordinarily located in the Top Object, which is the Blinker1 program. Naming the starting method Main is not mandatory but it does make reading Spin code much easier. I will discuss two statements in the Main method separately, but you must first realize that neither one will make much sense until you read the section below on the Output program. For right now, just continue reading, and I assure you that the material will become clearer as you finish reading all of it. Another point to remember is that several PUB sections may be in a Spin program and that each one contains its own method. How they are sequenced usually makes no difference except that if the Main method exists in a file, it should be first.

The first program instruction in the Blinker1's Main method is:

```
LED.Start(14, 6_000_000, 20)
```

The instruction is in an OO structure in which the reference name LED is to the left of the '.', which in OO terminology is known as the "member of" operator. I will simply refer to this term as "dot" from now on. The name "Start", which is directly to the right of the dot refers to a method named Start that is located and defined in the Output file. Remember that LED is the local name or reference to Output. The three values between the parentheses following the Start method name are arguments that are needed by Start to correctly perform its function. Right now, I will ignore their meaning, since it is not important for this particular discussion, but I will address it in the Output program discussion. Paraphrasing this instruction in OO terms would lead to the following:

> Execute the Start method, which is a member of the Output object, with these three specific data values.

The next instruction in the Main method is:

```
LED.Toggle(15, 4_000_000, 40)
```

By using the same logic as above, this instruction would translate to:

> Execute the Toggle method, which is a member of the Output object, with these three specific data values.

It is now time to examine the Output file Spin code so that you can fully understand what is happening with the foregoing instructions. Figure 4.8 is a screenshot of Output's Full Source code view. You may notice immediately that it does not contain nearly as much documentation and as many comments as does the Blinker1 code; nor should it, since it is a

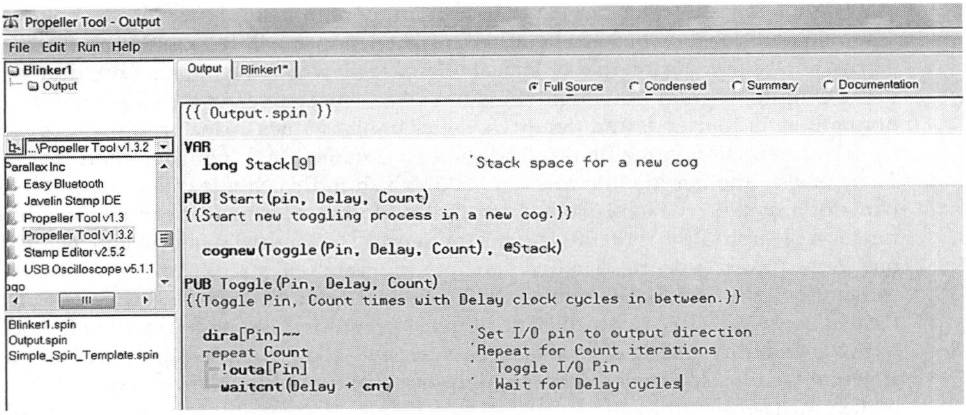

FIGURE 4.8 Output Full Source view.

component object. It does, however, contain a new section named VAR. The instruction contained in this section is:

```
long Stack[9]
```

This instruction reserves nine long data words in the hub's shared memory. Recall that the standard word for the Prop chip is 32 bits in length, which is equivalent to four bytes. Thus, nine longs reserves 36 bytes of space in the common hub RAM. This memory space is required to support the creation for a new cog's stack operating area. Without going into a detailed explanation, I will define a stack as simply a memory area in which an operating cog may store temporary data and an area where the Prop's firmware can logically access the cog as needed. A new cog is created when the Top Object, Blinker1, calls Output's Start method, which is the reason for reserving stack space.

Output also contains two public methods that I referred to earlier in the Blinker1 code discussion. The first method is named Start, and the second method is named Toggle. It is standard practice in Spin programming to have a Start method in a component object so that the Top Object may "kick it off" in a known and consistent manner. Output's Start method contains only one complex instruction, which causes a new cog to run that in turn executes the Toggle method. The Toggle method is Output's other public method. This new cog is automatically selected by the Spin interpreter from the first available cog, which in this case is cog #1. Cog #0 is always selected to start executing the Top Object code, and therefore, is already busy. The heart of the Output object is really the Toggle method. I have repeated the source code below so I could amplify the comments already contained in the code.

```
PUB Toggle(Pin, Delay, Count)
{{Toggle Pin, Count times with Delay clock cycles in between.}}
  dira[Pin]~~            ' Set I/O pin to output direction
  repeat Count           ' Repeat for Count iterations
    !outa[Pin]           ' Toggle I/O Pin
    waitcnt(Delay + cnt) ' Wait for Delay cycles
```

The first line of the code is also known as the method signature. It contains the public name of the method, which in this case, is Toggle, followed by any arguments it requires to function properly. There are three arguments—namely, Pin, Delay, and Count. Pin is an integer that represents the number of the GPIO pin that controls the LED. In our example, pins 14 and 15 are being used. Thus, it becomes immediately obvious that the Toggle method must be called twice to accommodate the two different pins.

The next argument is named Delay, and it is also an integer that represents a time delay in terms of actual clock cycles. There are two different delays used in the Blinker1 program, one with a 6_000_000 value and the other with a 4_000_000 value. I am sure you could not help but notice that underscores are used in lieu of commas for delimiters. I believe they are optional, but using underscores does help in avoiding the entry of an incorrect number of zeros. The actual delay time is dependent on the real Prop's operating clock speed. The 4_000_000 delay would represent an actual delay of ⁴⁄₁₂ seconds for a 12-MHz clock, which is the default rate used in the Spin interpreter, or equivalently, ⅓ of a second, which also equals 333.3 ms. Just remember: the bigger the number, the longer the delay.

The last argument is the Count, which represents the number of iterations or loops that the method performs before stopping. This sets the number of blinks, while Delay sets how long each blink lasts. The remaining instructions in the Toggle method are explained in Table 4.2.

I would like to go back to the Blinker1 code now that I have covered the Output source code. The two calls in Blinker1 should now make a little more sense to you. The first one is:

```
LED.Start(14, 6_000_000, 20)
```

The Start method in Output is called with pin 14 as the output, 6_000_000 as the delay in clock cycles, and 20 as the number of repeats or loops. Recall that Start also creates a new cog in which the Toggle method is executed.

The other Blinker1 call is:

```
LED.Toggle(15, 4_000_000, 40)
```

Instruction	Explanation
dira[Pin] ~~	Sets the GPIO pin whose value is Pin to be an output or source.
repeat Count	The beginning of a loop that repeats for Count value times.
!outa[Pin]	Toggles the value output from the pin, for example, 0 to 1 to 0 to 1 ... and so on. Note that the instruction is indented from the repeat instruction. This is how Spin determines which instructions are contained in the loop.
waitcnt(Delay+cnt)	This is the loop end because it is the last indented instruction. waitcnt delays Prop operations for the number of clock cycles in the parentheses. cnt is a global variable representing the current clock cycle number. It is not important per se since the Delay value is added to it, and the whole delay is really based upon the resulting amount.

TABLE 4.2 Toggle Spin Instructions

Here the Toggle method in Output is called directly with pin 15 as the output, 4_000_000 as the delay, and 40 as the number of loops. No new cog is created when Toggle is directly called so this instruction uses the existing cog that was started with Blinker1 or cog #0.

The complete BOE wired for the Blinker1 program is shown in Figure 4.9. It is quite simple—using only two LEDs, two resistors, and some jumper wires.

Porting to the Propeller QuickStart Board

I next wanted to demonstrate how easy it is to load the Blinker1 program onto a different Parallax development board. The board I selected was the QuickStart board, which is shown in Figure 4.10.

The QuickStart is a very low-cost Prop development board. It still comes equipped with a USB-to-Serial interface chip that makes it plug compatible with the Parallax PST. All you have to do is connect the QuickStart board to the PC running the PST software and load the Blinker1 program. The USB comm port on the QuickStart board should be automatically identified by the PC, which will allow you to download the program into the board's EEPROM. Figure 4.11 shows the board configured with a prototype solderless breadboard along with the LEDs connected in the same way as they were with the BOE.

I ran the Blinker1 program on the QuickStart board. In Figure 4.11, you can see that the board was powered by a 9-V battery. The program executed exactly the same as it did on the BOE.

Now that I have shown you how to develop some simple blinking programs, I would like to delve into some intricate Prop details.

Figure 4.9 BOE wired for the Blinker1 program.

FIGURE 4.10   Parallax Propeller QuickStart board.

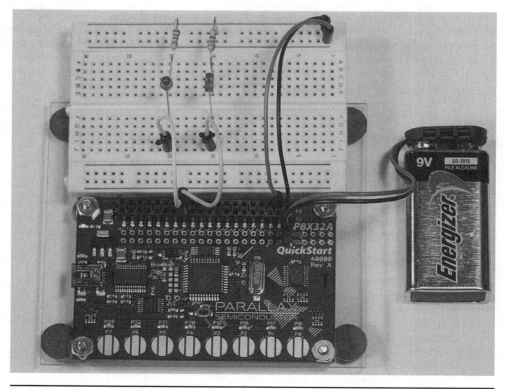

FIGURE 4.11   QuickStart prototype setup.

Clock Timing

This section is concerned with timing from both delay and duration aspects. Timing is a key feature in the real-time operations that the Prop performs. The duration of the LED blink in the Blinker1 program is based on the passage of a predesignated number of clock cycles that are assigned in the Delay variable. The total numbers of blinks are assigned by the Count variable, which also directly affects the total elapsed time that a specific LED blinks. The key instruction that implements the LED blink duration is:

```
waitcnt(Delay + cnt)
```

This instruction is within a loop in the Output's Toggle method. The number of loop iterations is controlled by the Count variable; hence, the total elapsed time for a blinking event is approximately computed by this formula:

$$\text{Total blink time} = (\text{Delay time}) * \text{Count}$$

Notice that I used the word approximately in describing the formula's accuracy because there is an extra brief delay from what is called overhead processing that involves executing loop instructions other than the actual delay instruction. This overhead, while typically very small, must be accounted for in extremely precise timing applications.

I conducted a very informal experiment to verify that the "Total blink time" formula was somewhat accurate. I simply timed the blinking duration of the LED controlled by pin 15. I did a set of four timing measurements and came up with a result of 13.4 seconds for the average total duration. But how does this compare with the theoretical time duration? To answer this question I had to compute the expected delay, as shown below:

```
Parameters:   Delay = 4,000,000
              Count = 40
Delay in seconds = 4,000,000 / 12,000,000 = 0.3333 seconds
     Total delay = 0.3333 * 40 = 13.33 seconds
        Measured = 13.40
      Calculated = 13.33
```

Less than 0.1 seconds between measured and calculated results isn't too bad for a crude experiment. However, you may be wondering where the 12,000,000 value came from in the "Delay in seconds" calculation. I will discuss this in the next section.

RC Oscillator Operations

Remember that at the beginning of this chapter I explained that the Prop chip had a variety of clock operational settings, including using an internal RC oscillator. It turns out that if you write a Spin program without specifying a particular clock configuration, the Spin compiler will automatically invoke what is called the "fast" RC oscillator (RCFAST) mode that runs at 12 MHz—this is where the 12,000,000 value comes from. This mode of operation is fine for initial prototyping and any non-time-critical application but is definitely not recommended for precision timing operations for all the reasons I have previously discussed. You should also note that the 12-MHz rating is what Parallax terms a nominal rating because it could range from 8 to 20 MHz depending upon a number of factors. I will soon discuss some programming approaches to mitigate the potentially adverse effects for this clock cycle variability.

There is one other RC clock mode that may be of some interest especially for extremely low-power applications. This mode is the slow RC clock mode in which the clock rate is only 20,000 Hz, which is almost 1000 times slower than the default RCFAST mode. Figure 4.12 shows the PST Full Source code view for the program named SlowBlinker1 that demonstrates this slow-speed operation. The Delay numbers have also been significantly reduced by a factor of 1000 due to the long times caused by using a slow clock speed.

The timing experiment was repeated for pin 15 and revealed a similar time of 13.36 seconds, which makes perfect sense, since the clock rate and delay were both scaled down by the same factor of 1000. A review of the SlowBlinker1 source code reveals that there is a new section named CON (which is short for CONSTANT). There is only one statement in the CON section:

```
_CLKMODE = RCSLOW
```

The name _CLKMODE is a built-in constant that the Spin software uses to set the desired clock mode. The name RCSLOW is simply equivalent to an integer that represents the clock mode. It does not matter what the actual number value is because Spin is programmed to respond to a given number to set a specific clock mode. Using this approach prevents an unfortunate programming practice known as magic numbers. To make this a bit clearer, let us suppose that the actual number for setting a slow RC clock mode is 8. Using this information would change the above clock mode expression to:

```
_CLKMODE = 8
```

This expression by itself makes no sense without the additional information that the number 8 actually represents the slow RC clock mode. That is why using 8 in the above

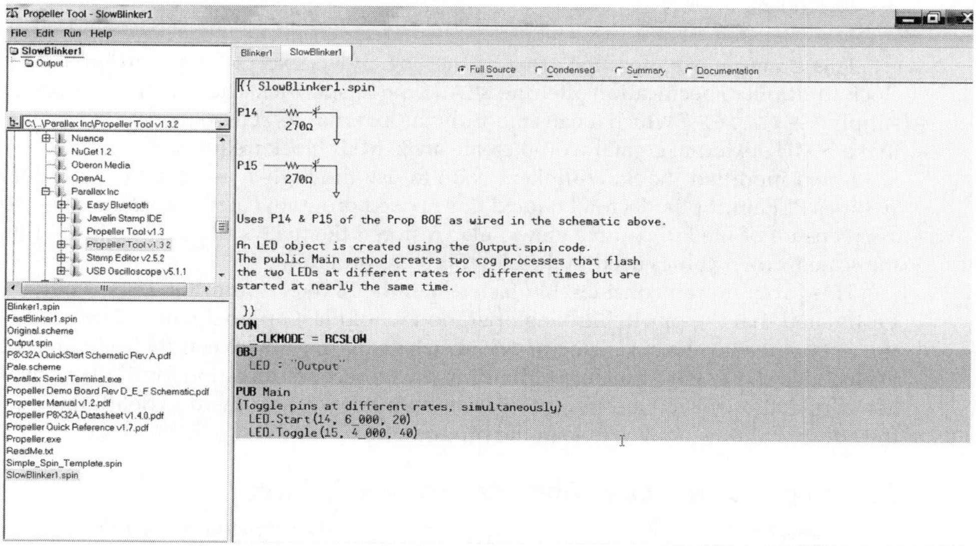

FIGURE 4.12   SlowBlinker1, Full Source view.

expression would be referred to as a "magic number"—it would take an act of magic to figure out what it represents. The programming practice that you should follow is to avoid using a magic number if at all possible, and if you do use one, ensure that you add a comment regarding what the number represents.

The RCSLOW clock mode is nominally rated for 20 kHz, but as was the case with the RCFAST mode, it has a potentially wide variation. This range goes from 13 kHz to 33 kHz, which would cause some serious issues if the time in your code was dependent on a preset clock cycle. In the following section, I will discuss how using a crystal oscillator can vastly improve the clock-cycle precision.

Crystal Oscillator Operations

Using the crystal oscillator is a simple matter of changing the clock mode. In this case, it involves two statements that need to be put into the CON section of the program Editor. The statements for a 5 MHz external crystal would be:

```
CON
    _CLKMODE = XTAL1
    _XINFREQ = 5_000_000
```

In this example, XTAL1 sets the clock mode for a crystal oscillator, and _XINFREQ specifies the external crystal resonant frequency that is connected to the X1 Prop pin. Recall that there is also a PLL frequency multiplier that may be used with an external crystal. In this example, no multiplicative factor is specified so the Prop clock frequency will be 5 MHz, the same as the external crystal. The next example shows you how to use a PLL multiplication factor.

```
CON
    _CLKMODE = XTAL1 + PLL16X
    _XINFREQ = 5_000_000
```

This example is almost the same as the one above, except for the addition of the PLL clock-multiplier specification after the XTAL1 mode specifier. The multiplier specification is simply " + PLL16X," which means multiply the external crystal frequency by 16. This means that a 5 MHz external crystal would create an 80 MHz clock frequency.

I next modified the SlowBlinker1 code to use the high-speed crystal oscillator with a 16 times PLL multiplier factor. I named this revised program FastBlinker1, and Figure 4.13 is a screenshot of the Full Source view. I also restored the delays to their original values from the scaled-down values used in the SlowBlinker1 program.

This program ran considerably faster than the Blinker1 program, as you might expect. I estimated that the pin 15 blinking operation would last approximately 2 seconds because the clock speedup was a factor of 6.5, which is the ratio between 80 MHz and 12 MHz. I calculated the 2-second result by dividing the 13-second operation for Blinker1 by 6.5. The actual operation was indeed around 2 seconds, but it was very hard to determine because of the additional time it takes to load the program from EEPROM into RAM.

Reducing Dependence on Absolute Clock-Cycle Times

In this section, I will show you how to get rid of the bothersome dependence on absolute clock-cycle timing when trying to set delay and duration times within your program. By

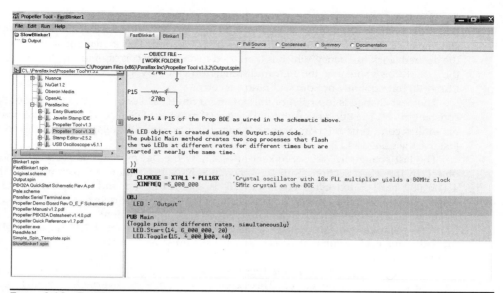

FIGURE 4.13   FastBlinker1 Full Source view.

following this approach, you will deal in units of milliseconds versus number of clock cycles, which will eliminate the requirement that you know the actual system clock speed in order to set actual delays and/or durations. I will demonstrate this approach by using a modified version of FastBlinker1 that I renamed PreciseBlinker1. Figure 4.14 is a screenshot of the Full Source view for PreciseBlinker1.

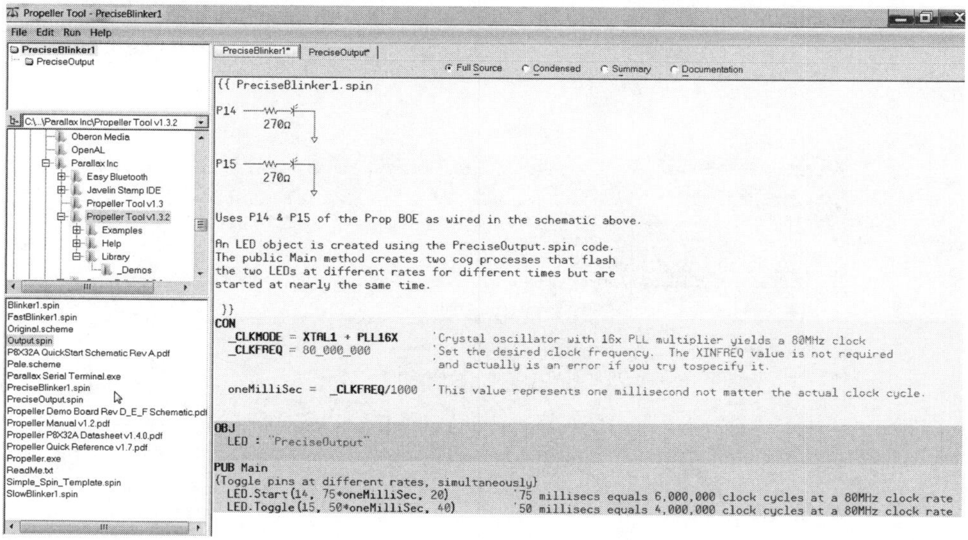

FIGURE 4.14   PreciseBlinker1 Full Source view.

One difference between the PreciseBlinker1 program and the FastBlinker1 program is that the statement _XINFREQ = 5_000_000 in the CON section has been removed, and the statement _CLKFREQ = 80_000_000 has been inserted. This statement specifies to Spin the desired clock frequency, which is then used to derive the corresponding _XINFREQ from the _CLKMODE value and the PLL multiplier. You can only specify either _XINFREQ or _CLKFREQ but not both, or Spin will return an error.

The next change is insertion of the statement oneMilliSec = _CLKFREQ/1000. The constant oneMilliSec now represents the total number of clock cycles that must pass in a one-millisecond time interval. You no longer have to explicitly relate clock cycles to time; just use the oneMilliSec constant.

The last change in PreciseBlinker1 affects how the Delay value is passed to the LED object. In the Main section, both Delay values have been changed from 6_000_000 and 4_000_000 hard-coded cycle counts to 75 × oneMilliSec and 50 × oneMilliSec, respectively. Now all you need do is focus on the desired time delay and not on counting clock cycles. This makes for much more pleasant code development.

A few changes were also made to the Output Spin object. Figure 4.15 is a screenshot of the Full Source view for that object, which I renamed as PreciseOutput to reflect the functional changes.

Two changes were made in the Toggle method. The first captured the system counter's value just before the loop started. This value is stored in a local variable named Time that must also now be declared in the Toggle's signature or top line. You can see that it is added after a vertical line delimiter. Time will hold a snapshot of the system counter value nearly at the instant the loop begins. The second change is in the loop in which the statement waitcnt(Time += Delay) is constantly evaluated until it is True; then the loop is ended. Recall that Delay's value is the total number of clock cycles that you desired to delay but expressed as a number of milliseconds times the number of clock cycles per millisecond. The "+=" expression in the waitcnt instruction is an "add to assignment" operator that instructs Spin to add the Delay value to the current Time value and store it back as a new Time value. It could also have been written as "Time = Time + Delay". This format however, is a bit more compact.

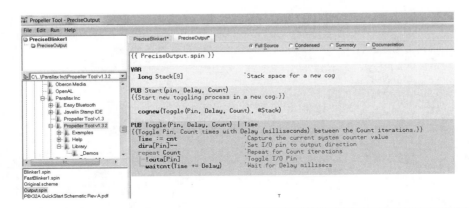

FIGURE 4.15 PreciseOutput Full Source view.

The PreciseBlinker1 program was downloaded into the BOE and run with exactly the same results as happened with the FastBlinker1 program, but PreciseBlinker1 deals with actual time rather than clock cycles.

I retested the PreciseBlinker1 program using a 40-MHz clock and got exactly the same results as with the 80-MHz version. This is precisely what I expected, thus proving that this approach makes dealing with delays and durations entirely independent of the actual clock speed.

Up to this point, I have demonstrated programs that toggle various GPIO pins at a 50% duty rate, which means that half the time they are on and the other half they are off. Viewing such a signal on an oscilloscope would reveal what is known as a square wave. Figure 4.16 is a screen capture of the PreciseBlinker1 output from pin 15.

Notice that the "on-time" or high portion of the signal trace is exactly 50 ms, as expected from the program. Of course, the "off-time" or low portion is also 50 ms, which makes the total waveform duration equal to 100 ms, or equivalently, 10 Hz.

I have included a photo of the oscilloscope that I used to measure the waveform for those readers who might be interested. Figure 4.17 shows a PicoScope model 3406B, which is a high performance PC oscilloscope. This instrument requires a PC to show the waveform because it has no organic display. The 3406B is a four-channel device capable of measuring signals up to 200 MHz with exceptional accuracy. I would urge readers to check out the Pico Technologies website to learn more about these highly capable and flexible instruments. However, be forewarned that they are not cheap; as the old adage goes, you get what you pay for. I am not disappointed in this instrument because it has performed flawlessly and enabled me to measure and record all the detailed waveforms that are shown in this book.

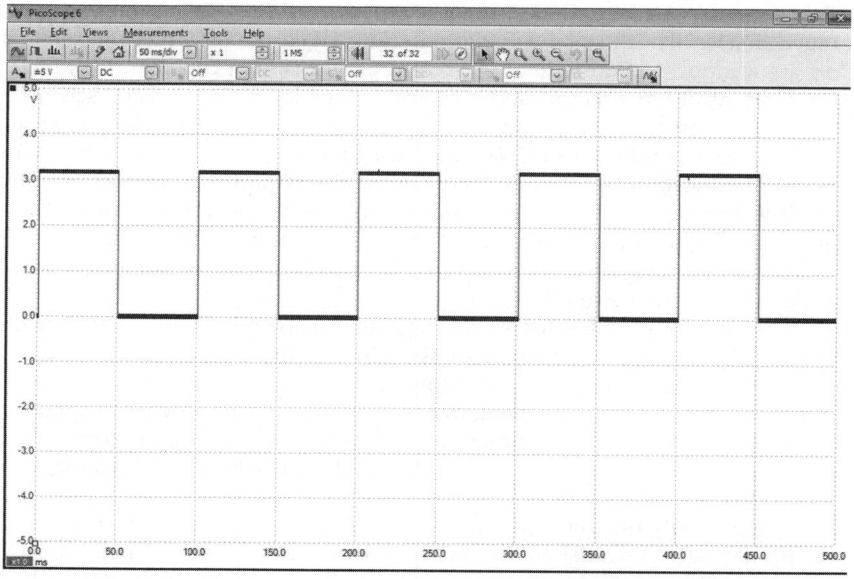

Figure 4.16 Oscilloscope real-time view of pin 15.

FIGURE 4.17 Picoscope Model 3406B USB oscilloscope.

Pulse-Width Modulation and Servo Example

This next example introduces *pulse-width modulation* (PWM) in which the pulse-on time is much shorter than the total waveform period. The pulse-on time is normally based upon a program input, while the total waveform period is kept constant. PWM is the technique used to control servo motors and is central to the Elev-8 flight-control scheme.

The demonstration program and Top Object are named 1 Center Servos and are part of the Parallax Learn Propeller Code tutorials. Two more component objects named Propeller Board of Education and PropBOE Servos are also required to run the Top Object. Running the Top Object creates a continuous waveform repeating at a 50-Hz rate with a pulse-on time of 1.5 ms. Figure 4.18 is a screenshot of the waveform measured by the USB oscilloscope connected to pin 14. The figure shows a 1.5-ms pulse that is repeating at 50 Hz, equivalent to a 20-ms waveform period.

The very simple 1 Center Servos program is shown in Figure 4.19. Looking at the code in the 1 Center Servos program shows that the author(s) was not terribly interested in thoroughly commenting or documenting the code. This lack of comments is a shame because it makes it hard for anyone using the code to understand how it is supposed to function. Admittedly, it is a very small program, but it still should have more comments. Having said that, I will proceed to discuss the program and show you how I made changes to demonstrate the PWM function.

There is only one method named Go in the program, and this is where the program starts executing. Yes, it should have been named Main, but as I mentioned earlier, it is a convention not a mandatory requirement. The first statement in the Go method is:

```
system.Clock(80_000_000)
```

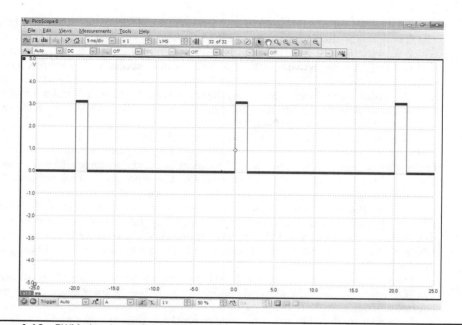

FIGURE 4.18   PWM signal waveform with a 1.5-ms pulse width and 50-Hz rate.

The Top Object is delegating clock configuration responsibilities to one of the component objects referenced by the reference name *system* whose file name is Propeller Board of Education. If you examine the source code for the Propeller Board of Education, you will find nearly the same code that I used in the FastBlinker1 program to set the clock. This delegation approach is useful, and it does minimize the code in the Top Object. It really boils down to your approach to coding, especially if there are multiple Top Objects to be used in a situation such as the Learning Tutorials where this code came from.

The next line in the Go method is:

```
servo.Set(14, 0)
```

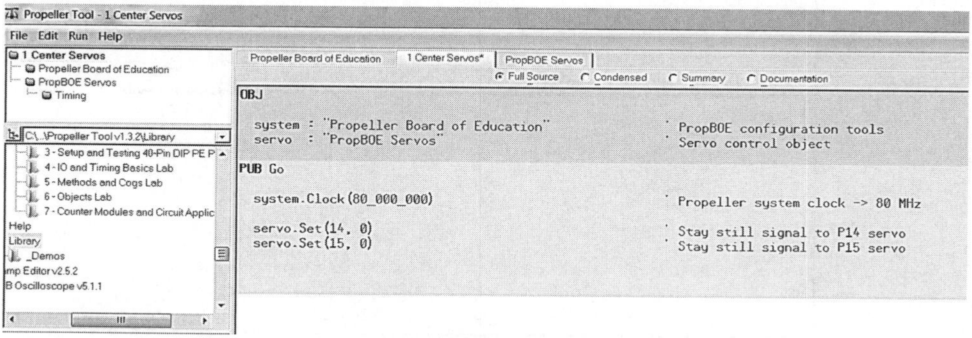

FIGURE 4.19   1 Center Servos Full Source view.

The local reference name is servo, and it represents the Spin file named PropBOE Servos. The Set method in this file controls the PWM waveform. The first argument is the GPIO pin, which is 14 in this case, and the second number is an offset. The offset represents a number that will range from −1000 to +1000. This offset can be thought of as a direct control of the pulse width where a 0 offset would be a 1.5 ms pulse, a −1000 would be a 0.5-ms pulse, and a +1000 would be a 2.5 ms pulse. Thus, the offset is really the number of microseconds that you want the pulse to deviate from the central 1.5-ms value. Any integer in the range from −1000 to +1000 will proportionally set the pulse width.

Most standard servos are designed to operate with this PWM technique. Sending a 2.5-ms pulse train to a servo designed to change angular position will cause it to rotate 90° in a clockwise direction from the neutral position. Similarly, sending a 0.5-ms pulse train will cause it to rotate 90° in a counterclockwise direction from the neutral position. Sending a 1.5-ms pulse train will keep the servo in its neutral position.

Sending the same type of pulses to a continuous rotation servo will cause angular velocity changes in which a 2.5-ms pulse would be the maximum speed in a clockwise direction, a 0.5-ms pulse would be the maximum speed in a counterclockwise direction, and 1.5 ms would be a speed of zero. Figure 4.20 shows a servo neutral positioning waveform that has a 1.5-ms pulse width.

I next connected a Hitec model HS-311 standard servo to the pin 14 servo connector on the BOE board. Figure 4.21 shows the BOE executing the 1 Center Servo program with the neutral position waveform being output.

I placed a small white arrow on the servo yoke to indicate the servo's neutral position, which, as you can see, is pointing to the top of the figure. I then changed the offset from 0 to +1000 and photographed the new servo position, as shown in Figure 4.22.

The servo has rotated 90° clockwise in response to the +1000 offset command placed in the Set command. The controlling waveform with a 2.5-ms pulse-width waveform is shown in Figure 4.23.

FIGURE 4.20   Servo neutral position waveform.

FIGURE 4.21   BOE controlling a Hitec servo in the neutral position.

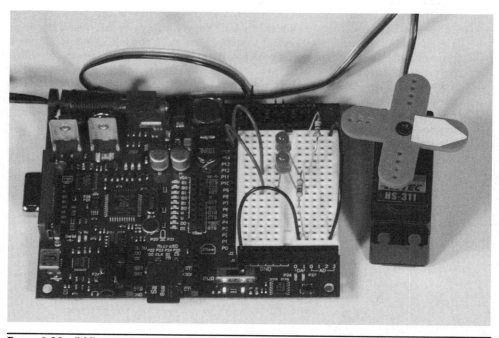

FIGURE 4.22   BOE controlling a Hitec servo in the maximum clockwise position.

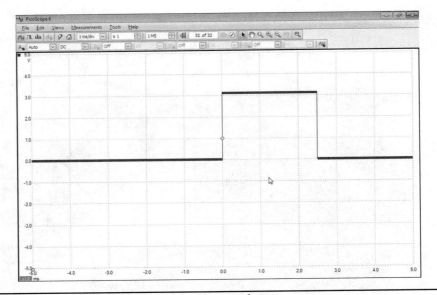

FIGURE 4.23   Servo maximum clockwise position waveform.

I tested the servo at other positions to confirm proper operation, but I will not take up book space to show servo positioning or waveform figures. You can be assured that the servo correctly responded to all the offset commands.

The PropBOE Servos Documentation View is available on this book's website www .mhprofessional.com/quadcopter to illustrate the complexity of this object. It also provides a guide for those readers who wish to experiment with the additional functions that are present in the code but not used in the foregoing example. In the example, Top Object calls only the Set method, which then calls the Start method, which in turn calls the Servos method. The pin and offset values are defined in the Set call, while some basic waveform parameters are defined in the Start method. The Start method also instantiates a new cog that will execute the Servos method where real action happens. I have replicated the Servos method code below, not so much to analyze it, but to show you how a real-time waveform generator may be created using the Spin language.

```
PRI servos | t, i, ch
  t := cnt
  repeat
    i := -1
    repeat until i == 13
      repeat ch from 0 to 1
        if ++i =< _servoCnt
          outa[_pinList[i]]~
          dira[_pinList[i]]~~
          spr[CTR + ch] := (%000100 << 26) & $FFFFFF00 |
_pinList[i]
          spr[FRQ + ch] := spr[PHS + ch] := 1
```

```
            pulse[i] += ((_pulseList[i] - pulse[i]) #> -_stepList[i]
<# _stepList[i])
            if ((_enableMask >> i) & 1)
               spr[PHS + ch] := -((pulse[i] #> -1000 <# 1000) * us +
center)
         waitcnt(t += frame)
      repeat until not lockset(lockID)
      longmove(@_pinlist, @pinList, 48)
      lockclr(lockID)
      waitcnt(t += cycleEnd)
```

You should first notice that the method begins with a PRI identifier, which is short for PRIVATE. This means that the Servos method is available only to other methods within the PropBOE Servos object and no other external object can call it. This restriction helps promote an important OO principle known as encapsulation. Objects should not reveal too much information on how they work internally so that unintentional changes are minimized.

The variable array _pinList[i] is set up to handle multiple servos operating simultaneously. This is a key advantage for the parallel processing that the Prop chip supports. It definitely is a great asset in this application.

There are two key statements in the above code that begin with the identifier spr. This is a Spin instruction that is short for *Special Purpose Register* (SPR), which allows you to indirectly access some specialized registers that are present in each cog. A register, for those unfamiliar with the term, is a named storage area where data can be read or written. I have included a list of a cog's SPRs taken from the *Propeller User's Manual* in Figure 4.24 because I think that you should be aware of them. They are an extremely valuable asset used in creating code that executes quickly and efficiently.

Table 2-15: Cog RAM Special Purpose Registers			
Name	**Index**	**Type**	**Description**
PAR	0	Read-Only	Boot Parameter
CNT	1	Read-Only	System Counter
INA	2	Read-Only	Input States for P31 - P0
INB	3	Read-Only	Input States for P63- P32[1]
OUTA	4	Read/Write	Output States for P31 - P0
OUTB	5	Read/Write	Output States for P63 – P32[1]
DIRA	6	Read/Write	Direction States for P31 - P0
DIRB	7	Read/Write	Direction States for P63 - P32[1]
CTRA	8	Read/Write	Counter A Control
CTRB	9	Read/Write	Counter B Control
FRQA	10	Read/Write	Counter A Frequency
FRQB	11	Read/Write	Counter B Frequency
PHSA	12	Read/Write	Counter A Phase
PHSB	13	Read/Write	Counter B Phase
VCFG	14	Read/Write	Video Configuration
VSCL	15	Read/Write	Video Scale

Note 1: Reserved for future use

FIGURE 4.24 Cog Special Purpose Registers (SPR).

I would like to point out that the above code is so efficient that it can replace the need to use assembly language code, which is ordinarily used to achieve the performance necessary for this application. However, do not be dismayed. I will discuss assembly language in Chapter 5, since it supports the demonstration program used to control an Elev-8 test motor.

Summary

The chapter began with an introduction to the Parallax Propeller chip's unique architecture. The cores or cogs were shown to be highly flexible computing elements capable of performing parallel tasks to efficiently execute application instructions. I also discussed the hub that coordinates cog activity and the highly flexible clock circuits that support the Prop chip.

The *Propeller Spin Tool* (PST) discussion included a demonstration of how easy it was to create, load, and execute programs using the PST and a Prop development board. I showed you the Propeller Board of Education (BOE) that I used for software development. It includes a USB-to-serial interface chip that makes connecting the BOE to a PC running the PST, an effortless task.

I next went through a series of LED blinker programs that demonstrated basic Spin programming as well as some basic object-oriented (OO) techniques. The PST makes it very easy to get started with Spin programming. Of course, continued study and practice is the only way to develop solid software development skills.

One of the LED blinker programs was next transferred to a Parallax QuickStart development board. I did this to show that transferring a Spin program developed on the BOE to another Parallax development board raised no issues whatsoever.

Next came a clock-timing discussion in which I went through the various system clock modes and explained the pros and cons for each one. I recommended that the external crystal-controlled oscillator be used for your application program development. This clock mode is accurate, fast, and readily available on most of the Parallax Prop development boards.

I next showed you how to use actual time values in lieu of clock cycles when creating your programs. Using time values expressed in milliseconds is far superior to using clock cycles. This technique makes time values used in your program independent of the actual running system clock.

A comprehensive *pulse-width modulation* (PWM) discussion followed the clock tutorial. This PWM introduction lays the foundation for a proper understanding of the technology integral to servo operation. Servo technology is used extensively in radio-controlled flight systems as well as for the Elev-8 flight system. I used several programs to demonstrate how basic servo-control algorithms function.

The chapter concluded with an introduction to the cog special purpose registers (SPR). These registers are the key to creating efficient and very fast application programs.

The next chapter examines some of the Elev-8 critical components, including motors, ESCs, and propellers (the ones that whirl, not the chip kind).

CHAPTER 5

Quadcopter Propulsors

Introduction

As the title indicates, this chapter's discussion will encompass the three elements that produce the thrust that propels the quadcopter into the air: the propeller, the motor, and the *electronic speed controller* (ESC). Each one depends on the other two; and without each doing its part, there will be no thrust created, and the quadcopter will remain stationary on the ground. Each element will be discussed separately; however, I will try to show how they are each tied to one another and will identify their limitations and the constraints they impose on each other.

The chapter will also present a demonstration of two programs that will allow you to further explore your current propellers and motor/ESC combinations, along with possible future propeller upgrades.

I will begin by discussing the motor, an element critical to quadcopter performance and one most people are familiar with.

Motors

Direct current (DC) motors are nearly universally used in R/C aircraft, helicopters, and multirotor craft. The two main DC motor types are *brushed* (BDC) and *brushless* (BLDC). Brushless motors are preferred for use in quadcopters because they do not use carbon brushes, which makes them much easier to maintain. They also rotate at very high speeds, as compared to brushed motors, and produce less electrical noise. One of the Elev-8 BLDC motors, a model A2212/13T 1000KV, is shown in Figure 5.1.

The 1000KV in the model name means that the motor is designed to rotate at 1000 r/min per volt applied to the motor. Thus, an ESC that is powered by a fully charged three-cell LiPo battery producing over 12 volts could theoretically rotate the motor at a maximum of 12,000 r/min. In reality, the maximum rotation is about 7000 r/min for the Elev-8, since that is about as fast as the Slo-Flyer propellers can efficiently rotate. In addition, a motor turning a propeller load will typically have an approximately 600-r/min-per-volt rating, which equates nicely with the maximum 7000 r/min at roughly a 12-V supply voltage.

BLDCs used in the Elev-8 are also unique in that the rotors are on the motor exterior, while the stators are fixed on the inside of the motor. This type of motor is also known as an

FIGURE 5.1   Elev-8 BLDC motor A2212/13T 1000KV.

outrunner because the rotors are on the motor's exterior. Naturally, traditional rotors that are inside the motor casing are known as inrunners. Outrunner motors have rotors that use *permanent magnets* (PMs), as can be seen in Figure 5.2.

The PMs used in the Elev-8 motors are made of Neodym, which has the chemical nomenclature of NdFeB. The elements that make up Neodym are neodymium (Nd), iron (Fe), and boron (B). Neodymium is classified as a rare earth element and is the key magnet constituent. It has a very high magnetic permanence, which means it can be permanently magnetized and maintain a very strong field.

There are 14 PM bars mounted on the inside of the rotor, as may be seen in Figure 5.2. The magnets are mounted with alternating north/south magnetic field orientations in order to match the rotating-stator electromagnetic field.

FIGURE 5.2   BLDC 14-pole PM rotor.

FIGURE 5.3   BLDC 12-pole wire-wound stator.

The other main motor part is the stator, which is made up of wire-wound coils that wrap around an iron core. Figure 5.3 is a photo of a 12-pole stator similar to the ones used in the Elev-8 motors.

It is important to note that there are 14 poles on the rotor and 12 poles on the stator. The difference in pole numbers is crucial to starting and maintaining motor rotations. The rotor would otherwise lock up and not rotate if the number of poles on the rotor and stator were equal. Figure 5.4 is an end view of an assembled A2212/13T motor that clearly shows the displacement between rotor and stator poles that ensures the motor will rotate.

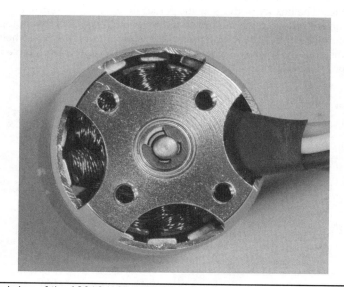

FIGURE 5.4   End view of the A2212/13T 1000KV motor.

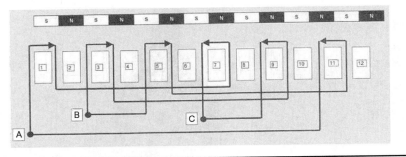

Figure 5.5   BLDC stator wiring diagram.

Also critical to motor rotation is the electrical supply that powers the stator. It is effectively a three-phase power supply that is created by an ESC, which in turn is connected to the stator. A simplified wiring diagram is shown in Figure 5.5.

This figure reflects a portion of the wiring that illustrates how the three leads from the ESC are connected to the stator coils. The A-, B-, and C-connection points alternately conduct current in such a way that the stator creates a rotating magnetic field. If you trace the wire from the A terminal, you can see that it wraps around stator pole 1 in a clockwise direction, and also, through stator pole 11 in a counterclockwise direction. The net effect of opposite current flow is to create opposite magnetic poles at each of the physical stator poles. These electromagnetic poles are close to the PM poles on the rotor assembly. This whole interaction causes the rotor to move while the electromagnetic poles are moved through the current switching that is happening through the A, B, and C terminals. I realize this is all a bit confusing, but the whole action is carefully orchestrated by the programming contained in the ESC, which is designed to work with the fixed physical dimensions of the stator and rotor poles. The beauty of this scheme is that the ESC need control only the rate at which it sends current pulses to the motor, which then directly control the motor's rotational speed.

A simplified physical wiring diagram is shown in Figure 5.6 to help clarify the way in which the stator is wound. To help clarify the BLDC operation, I have included Figure 5.7, which is a snapshot of an animation showing a BLDC in action.

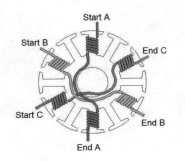

Figure 5.6   Simplified physical stator wiring diagram.

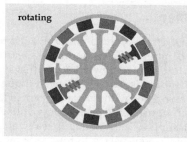

FIGURE 5.7   Stop-action screenshot of a rotating, animated BLDC.

I would also suggest going to the website http://www.aerodesign.de/peter/2001/LRK350/LRK_in_action.gif, if you wish to see the animation of a BLDC in action.

The stator coils themselves are internally connected in either a Wye or Delta configuration. These two configurations are shown in Figure 5.8.

It makes little difference to the user if the motor is wired in a Wye or Delta configuration. The real difference comes from the fact that motors wired as a Delta can operate at higher speeds and voltages while using somewhat less operating current then those wired as a Wye. The Wye-wound motors are a bit slower but produce more torque at the expense of higher operating currents. Of course, higher currents mean higher operating temperatures, which is something to avoid in these BLDC motors.

Heat is always an issue with motors, especially with ones that use strong Neodym PMs. There is a phenomenon known as the Curie Point, where a magnet can lose its magnetism due to excessive heat. It can be as low as 80° C for an ultra-strong rare-earth magnet like the Neodym. This is still a very high temperature as compared to a normal ambient temperature, but you should realize that there is plenty of current coursing through these motors when they are operating at full tilt. I would estimate that operating at 100% capacity for 20 minutes would likely raise the internal motor temperatures to this region. It would be a real shame to demagnetize the motors, especially if the quadcopter was still flying. Of course, the motors are irreversibly ruined once the PMs are demagnetized. The procedure to adhere to is to occasionally back off and operate at a slower r/min to let the motors cool.

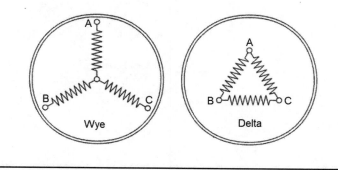

FIGURE 5.8   Wye and Delta configurations.

Electronic Speed Controller

The primary purpose of an *electronic speed controller* (ESC) is to supply power to a motor that is proportional to its control input, which is normally a servo-type signal. The ESC supplies power to the motor via a three-phase power supply that was first discussed in the above motor section. The power supply is strictly DC, even though I used the descriptor "three-phase," which is normally associated with *alternating current* (AC) motors. The power-supply voltage varies only between zero and the peak battery voltage and never goes negative as it would with an AC power supply. The phasing is really about the current-pulse sequence that is delivered to the motor and causes it to rotate. (Some figures that are shown later in this section should help clarify the phase concept.)

Refer to the ESC block diagram shown in Figure 5.9 as you read the description of how an ESC functions.

The heart of modern ESCs used in quadcopter projects is the Atmel ATmega8L *microcontroller unit* (MCU), which is a flash-memory-based, 8-bit microprocessor with some peripheral-control components. Table 5.1 shows some of the ATmega8L key specifications.

These specifications reveal a very capable controller that can easily handle the demanding real-time task of converting servo signals to their equivalent three-phase power pulses. You should note the three *pulse-width modulation* (PWM) channels that are included in the controller circuits because they are important components in generating the three-phase power control signals.

The essence of an ESC is rather simple: it just chops up the battery supply voltage and sends these power pulses to the motor coils in a sequence that generates a rotating electromagnetic field in the stator. The MCU creates gate control signals that are sent via traces labeled as A, B, and C in the block diagram to the MOSFET switches. The three-phase control signals are next sent to gate circuits that control a series of MOSFETs that, in turn, switch on the raw battery power. The switched power is then fed to the motor via the A, B,

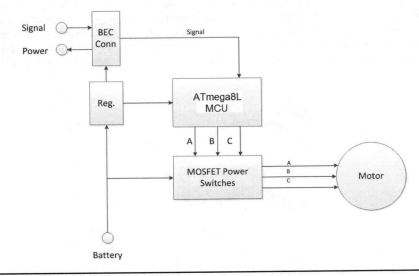

Figure 5.9   ESC block diagram.

Feature	Description
Flash memory	8 kB capable of in-system programming
RAM	512 B
SRAM	2 kB (holds configuration data)
Max clock speed	8 MHz (internal R/C oscillator)
Analog-to-digital conversion	8 channels with 10 bit accuracy
Operating voltage/current	2.7 to 5.5 V/1 mA (idle) to 3.6 mA (active)
Interrupts	2 external
Counters/timers	Two 8-bit Timer/Counter, One 16-bit Timer/Counter
PWM	3 PWM channels
Serial interfaces	1 USART, 1 SPI, 1 I²C
Real-time counter	1 System counter with separate oscillator

TABLE 5.1 Key ATmega8L Specifications

and C power leads. You can think of the MOSFETs as a series of high-speed, solid-state power switches. They are connected in parallel to be able to handle the high currents associated with quadcopter operations. I have included a detailed ESC schematic on this book's website, www.mhprofessional.com/quadcopter, for those readers who are interested in looking at an actual circuit for a typical ESC. This schematic shows the circuitry for a Tower Pro series ESC rated for a peak 25-A load.

Figure 5.10 is a picture of a 25-A HobbyKing ESC in which you can see the three motor leads coming from the left side and the cable for the *battery eliminator circuit* (BEC) and battery power leads coming from the right side of the unit.

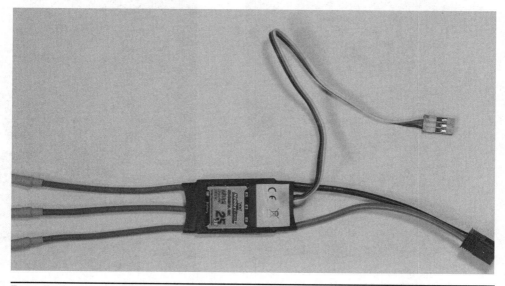

FIGURE 5.10 A 25-A ESC.

Most ESCs, especially those made in China, are covered in a large heat shrink tube that prevents inadvertent short circuits and offers some physical protection to the PCB components. This type of covering, while very inexpensive, does not readily dissipate internal heat or allow easy access to onboard components. I did remove the heat shrink tube from a 25-A Turnigy ESC to show you the components that are hidden beneath it. The front side of this ESC is shown on the left side of Figure 5.11, while another exposed ESC photo that I gathered from an R/C website is shown on the right side of the figure.

A comparison of the two boards should reveal that despite being from different sources they are almost identical, at least as far as component type and placement. I believe that Chinese manufactured ESCs are produced at only a few factories and then sold under a variety of brand names. It is very much a commodity market, and you are quite likely to receive exactly the same ESC for a particular amperage rating, no matter the branding. There is another, darker side to this type of marketing that you should also know about. Some unscrupulous marketing companies buy rejected ESCs from reputable distributors, rebrand them, and then sell them as fully functional ESCs. I would strongly suggest that you purchase only from reputable dealers and distributors and avoid the secondary market, where you may get a much cheaper price, but also much lower quality. Remember that the ESCs supply the power to the motors that keep your expensive quadcopter aloft.

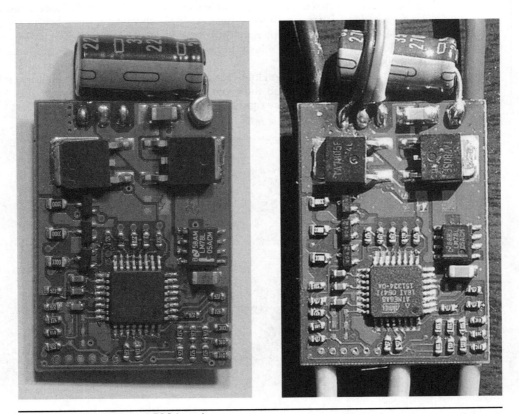

FIGURE 5.11 Two exposed ESC boards.

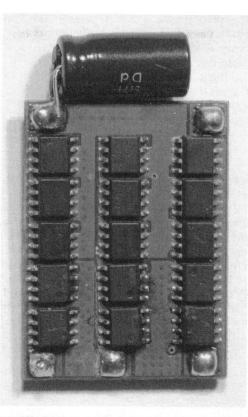

FIGURE 5.12 25-A Turnigy MOSFET columns.

Figure 5.12 shows the back of the 25-A Turnigy board, on which you can easily see the *metal-oxide-semiconductor field-effect transistor* (MOSFET) columns. Each column powers one of the A, B, or C leads and contains five power MOSFETs that are connected in parallel so that each one can handle up to a maximum of 5 amperes of current. The power MOSFET used on this board is a generic p-channel type with a model number of 4407. As used here, the term *generic* means that a variety of manufacturers can supply this chip as long as it meets the functional specifications listed in Table 5.2.

Specification	Description
Maximum voltage	−30 V
Maximum current	15 A
Package	SOP-8

TABLE 5.2 4407 Power MOSFET Specifications

Notice that the current rating is 15 A. This means it is very conservatively rated for this application, in which the maximum current should not be more than 5 A. The maximum voltage is also conservative, since the real battery voltage will likely remain below 13 V. The key take away from this specification review is that the 30-A Elev-8 ESCs should easily handle any normal flight operations without being overstressed or overheated.

Next, I will discuss the waveforms associated with ESC operations, which will help you understand how the ESC functions. I am postponing the BEC discussion until later in this chapter in order to establish a good foundation to understand what happens with the BEC circuits.

ESC Waveforms

An oscilloscope captures the waveform of typical signals that are sent from the MCU to the MOSFETs, as shown in Figure 5.13.

A gate control signal will turn on one of the MOSFET columns when it has a positive voltage. You can see from the figure that Gate A is on for 1 ms, then turns off, and Gate B immediately turns on. Gate B stays on for 1 ms, shuts off, and Gate C turns on for 1 ms. The whole process repeats every 3 ms, or approximately 333 Hz. This is the origin of the three-phase operation that I mentioned earlier in the chapter. Remember, a positive gate voltage will turn on the MOSFETs to which it is connected. The MOSFETs will then conduct and allow current to flow through their connected motor coils, thus causing the rotating electromagnetic field. It is a fairly simple but somewhat elegant scheme that creates a pseudo three-phase motor whose speed and torque can be closely controlled by the MCU in the ESC in response to external servo-control signals.

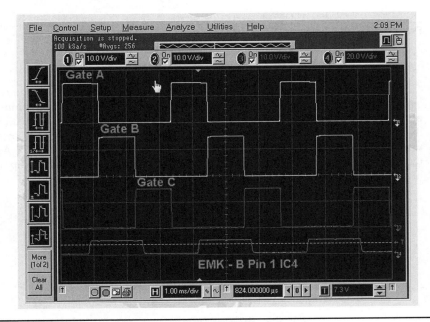

FIGURE 5.13   Gate control signals from MCU to MOSFETs.

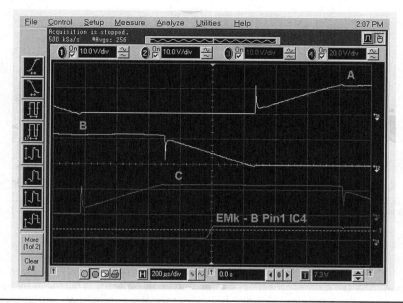

FIGURE 5.14 Phase A, B, and C voltage waveforms.

Figure 5.14 is a snapshot of the voltage waveforms on the A, B, and C leads that are connected to the motor. You should notice that the voltage ramps up from zero to the battery supply voltage and then back down to a zero level. The ramping is due to the selected configuration settings that you can program into the ESC. Also, notice that there are some sharp peaks present in the waveforms that are due to switching transients. These peaks are normally not a concern because they are very brief and the inductive nature of the motor circuit will tend to smooth out these transients.

Now that I've finished the ESC theory discussion (except for the BEC information that will be covered in a later section), I will show you an original experiment that demonstrates some interesting propeller, motor, and ESC interactions.

Propeller, Motor, and ESC Experiment

WARNING: This is a potentially dangerous experiment because it involves a sharp, hard-plastic propeller that is spinning at very high speeds. This propeller is completely invisible when rotating at high speeds, and users are at risk of serious injury or worse if they inadvertently come in contact with the spinning propeller. This experiment should not be conducted by unsupervised children or others not fully capable and aware of the inherent dangers. I do suggest some ideas to mitigate the potential hazards, but I strongly suggest you simply use my reported results and avoid repeating the experiment if you feel uncomfortable in doing so.

This experiment places one of the Elev-8 motors at one end of a miniature seesaw or "teeter-totter," with the other end supported by a force scale. Figure 5.15 is a sketch of the experimental setup.

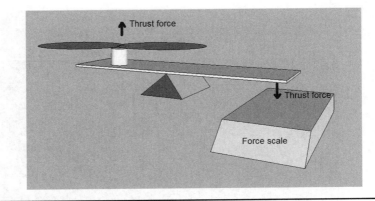
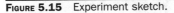

FIGURE 5.15   Experiment sketch.

The experiment is based on the balancing of forces in which the upward force or thrust created by the motor/propeller combination on the left side is balanced by the force-balance gauge on the right side. Figure 5.16 is a picture of the miniature seesaw that I built from a ¼-in thick acrylic beam to pivot on a wood base.

The seesaw base is made of maple wood because it is strong but easily shaped. I also placed a nail through a hole drilled through the acrylic beam to form the pivot. It turned out that the thrust forces were so strong that the beam actually deformed a bit, but I don't believe it affected the force readings.

An A2212/13T motor is being driven by a 25-A Turnigy ESC that, in turn, is controlled by the BOE running a program named ESC_Motor_Control_Demo. I will discuss this program in depth in a later section, but for now I will describe only how it functions in this experiment.

The test circuit shown in Figure 5.17 uses a LiPo 3S battery rated at 5000 mAh with a 40–80 C discharge capacity as the ESC power source. Needless to say, the battery is more than sufficient to power this motor/propeller combination. Not shown in the diagram is the laptop that the user needs to connect to the BOE in order to control the motor speed.

FIGURE 5.16   Minature seesaw.

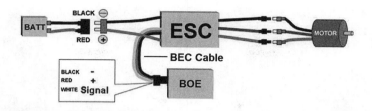

FIGURE 5.17 Experimental test-circuit diagram.

The actual test circuit is shown in Figure 5.18 with all the components interconnected. The ESC BEC cable is connected to the BOE solderless breadboard with the BOE's pin 14 and ground connected to it. No other control connections are needed for this setup.

The complete test setup, including the laptop that you use to control the motor speed, is shown in Figure 5.19. Notice that the seesaw base is C-clamped to the tabletop. The vibrations are fairly severe, especially at higher speeds. Not clamping the base would cause the seesaw literally to start flying, which is definitely not a desirable experimental outcome.

An optical tachometer is the last item needed for this experiment. Figure 5.20 shows the optical tachometer that I used. This tachometer is a clever little device with which you can directly measure the propeller's rotational speed by simply pointing it at the propeller from

FIGURE 5.18 Actual test circuit.

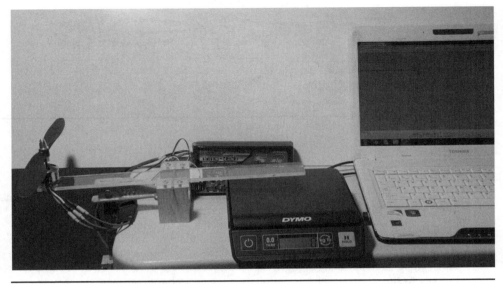

FIGURE 5.19   Complete test setup.

a distance of 4 to 6 in. A sensitive photo cell at the top of the unit detects light reflections from the propeller. Notice that the unit is showing 3600 r/min in the figure, which is due to the flicker rate of the lights used in the photography setup. Pressing the button on the front of the unit a second time will switch it from a two-bladed to a three-bladed measurement mode—definitely a handy tool to possess.

Running the Experiment

Before the experiment can start, the demo program mentioned above must be loaded into the BOE's EEPROM. The test circuit must also be connected, as shown in Figure 5.17, with

FIGURE 5.20   Optical tachometer.

the BOE's pin 14 connected to the BEC's signal lead (usually white or orange) and the ground leads connected. I attached an oscilloscope lead to the signal lead in order to measure the waveforms being created by the BOE. I report some key timing measurements in the results section below.

Note that the BEC's power lead (usually red) is left unconnected. I did not plug the BEC connector directly into the BOE's P14 servo pins because doing so would have tied the BEC's power into the BOE's power, which is not a desirable configuration.

A USB serial cable connects a laptop running the Parallax Serial Terminal (PSerT) program to the BOE. Entering data using the PSerT program is the way you control motor speed. You just need to enter a single number from 0 to 8, which represent power levels from 0% to 100% respectively. The eight steps mean that increasing each number by one is equivalent to increasing the power level by 12.5%. It turned out that the maximum level I could reach was 5, or 62.5% of maximum power. Anything above 5 simply created way too much vibration and energy in the experimental setup to the point where I felt unsafe in continuing the operations. However, I did test the motor up to level 8 without a propeller attached. This was simply to check that the program functioned as expected and to measure some waveform timing parameters.

Caution: I highly recommend that you place some sturdy barriers around the area where the propeller is spinning to prevent anyone from inadvertently touching or coming in contact with the spinning propeller. Remember that the propeller is completely invisible when it is spinning. Touching a spinning propeller will cause serious injury—no question about it! You might want to use some large pieces of foam board attached to chairs to fashion a reasonable barrier system. I am very cautious around propellers, whether they are the small ones discussed here or the large ones, such as the type I use when flying a light plane. Coming in contact with one of those when it is spinning would be the last thing you do on this earth.

Figure 5.21 is a screenshot of the PSerT communications screen connected to the BOE that is running the demo program. You must press the space bar in order to start entering

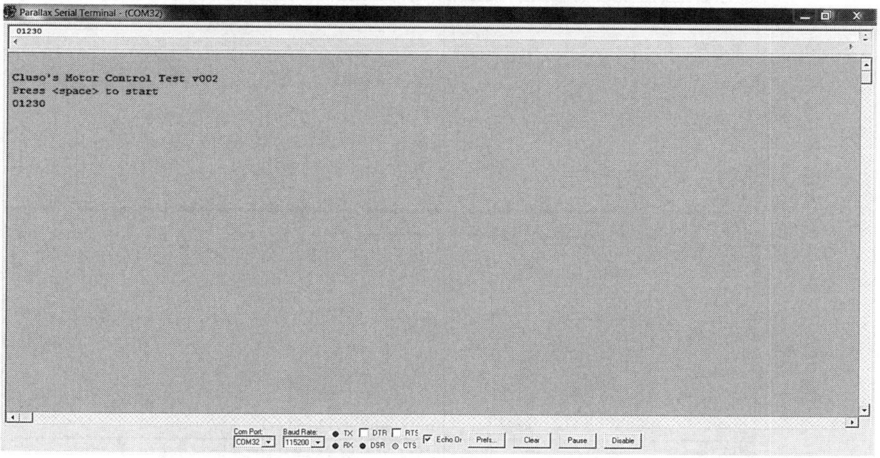

Figure 5.21   Screenshot of the PSerT connected to the BOE running the demo program.

numbers, as you can see in the figure. In this screenshot, I started with 0 and then entered 1, 2, 3, and 0. The demo program is designed to start with 0% power so that you don't get startled when you press the space bar. That is all that is needed to run this experiment, and in the next section, I will both show and discuss the results.

Experimental Results

Timing measurements were based on the waveform diagram shown in Figure 5.22. The pulse width is T1, and the spacing between pulses (period) is T2, as shown in the figure.

In Table 5.3, you see the combined results of running the motor with and without the propeller. Obviously, no thrust readings were possible without the propeller attached, but it was possible to capture all the timing readings. Some results, such as those for power, are computed from other table values.

The x's in the Table 5.3 indicate which measurements were not taken because of excessive vibrations or because the limits of the measuring instruments were exceeded, or both. Nonetheless, I was able to record a reasonable set of performance figures for low- to medium-speed operation. Figure 5.23 shows a chart of thrust versus power level, which appears to be a reasonable curve based on my research.

Figure 5.24 shows a chart of r/min versus thrust, which is a common method used to assess the performance of a propeller and is discussed in a following section. I also added an additional data point that I estimated from the trend of the actual data points. This estimate reveals that 860 grams of thrust will be generated when the propeller is spinning at 7000 r/min.

Figure 5.25, which shows the power consumed versus propeller r/min, is the final chart that I will show. I added an additional data point that I estimated from the trend of the actual data points. This estimate reveals that 150 watts (W) of power will be required to spin the

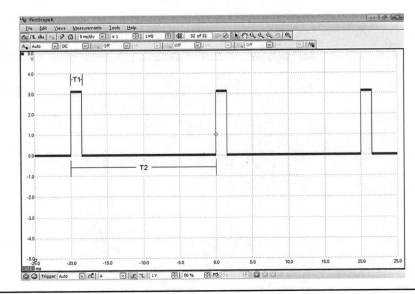

FIGURE 5.22   Waveform measurements diagram.

Power Setting	T1 (ms)	T2 (ms)	r/min	Thrust (grams)	Amperes (w/prop)	Voltage (V DC)	Power (W)
0	1.000	21.02	0	0	0.020	12.56	.25
1	1.125	21.12	1140	50*	0.23	12.56	3.01
2	1.250	21.22	3450	168	1.76	12.53	20.05
3	1.375	21.38	4710	343	4.08	12.44	50.76
4	1.500	21.48	6030	631	8.35	12.35	103.12
5	1.625	21.58	7020	x	x	x	x
6	1.750	21.74	x	x	x	x	x
7	1.875	21.84	x	x	x	x	x
8	2.000	22.00	x	x	x	x	x

*This is an estimate because the motor/prop did not create enough thrust to press on the force scale. It is based on the weight of the motor/prop combination.

TABLE 5.3   Experimental Results

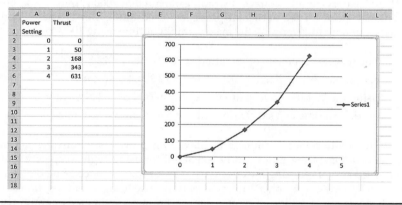

FIGURE 5.23   Chart of thrust versus power setting.

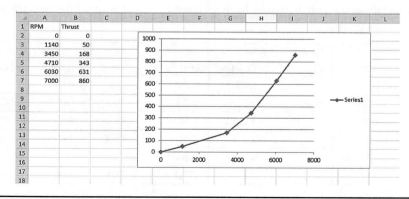

FIGURE 5.24   Chart of thrust versus propeller r/min.

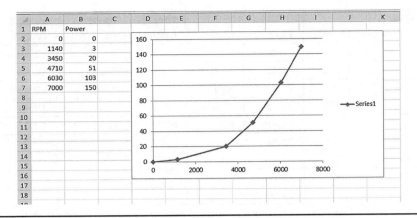

Figure 5.25 Chart of power versus propeller r/min.

propeller at 7000 r/min. 150 W would mean that approximately 12.3 A would be needed at a 12.2-V battery-supply voltage. I would estimate that 7000 r/min would be the maximum rotational speed for the particular propellers used in the Elev-8. The following calculation shows you how to determine the maximum flight time using 100% power:

One motor/propeller combination at 7000 r/min = 12.2 A

Four motor/propellers at 7000 r/min = 12.2 × 4 = 48.8 A

Using the 3S LiPo battery discussed in Chapter 3 = 4200 mAh = 4.2 Ah

Maximum time at 7000 r/min = 4.2/48.8 × 60 min = 5.16 min

Wow! Only about 5 minutes at maximum power is a startling fact for quadcopter operations. Operating times can be extended by using higher-capacity batteries, but that comes at a cost of reducing the effective payload capacity, since using bigger batteries means heavier batteries. The other preferred way of extending operating times is to operate at much lower r/min settings. I created Table 5.4 to show estimated operating times and thrust versus r/min for the battery described above.

Table 5.4 shows how the propeller speed affects both time and the creation of thrust. It is definitely a tradeoff that you have to consider constantly while you are operating the

r/min	Power (W)	Voltage (V DC)	Amperage (A)	Time (min)	Thrust (grams)	Total Thrust (grams)
0	0	12.56	0.02	n/a	n/a	n/a
1140	3	12.56	0.24	263.8	50	200
3450	20	12.53	1.60	39.5	168	672
4710	51	12.44	4.10	15.4	343	1372
6030	103	12.35	8.34	7.6	631	2524
7000	150	12.20	12.3	5.1	860	3440

Table 5.4 Thrust and Time versus r/min

quadcopter. The basic Elev-8 weighs approximately 1400 grams (including the battery), which means it will not even lift off unless the propellers are spinning at least 4800 r/min. That translates to about a 15-min operating cycle, which nicely matches my actual operating experience. If you add several hundred grams of payload, you should not expect any more than 10 to 12 min of flight time, assuming you do not operate at maximum power.

Now I want to discuss the timing values of the ESC, as shown in Table 5.3. The T1 column shows the control-pulse width for the various power settings. The pulse ranges from 1.0 ms for zero power to 2.0 ms for 100% power. It is linear and proportional, which means that a 1.5-ms pulse width would represent 50% power. It is useful to know this relationship, especially if you decide to create your own flight-control board. The pulse period goes from 21.02 to 22.00 ms because of the way the demo program creates the pulse train. This means that the frequency ranges from 45 Hz (22 ms) to approximately 47.6 Hz (21.02 ms). Although it does not matter at all for this demo program, it might affect flight-control responsiveness if this program code were to be incorporated into a quadcopter flight-control program. The whole matter of ESC pulse-train frequency is a bit controversial, as there are a number of quadcopter designers who insist that ESCs should have a 400-Hz operating frequency versus the standard 50 Hz. Most everyone in the R/C field agrees that the 50 Hz is quite adequate for the typical, fixed-wing aircraft. However, some insist that it is not nearly responsive enough to match the desirable quadcopter flight-response characteristics. I am still unsure about this claim, although I am experimenting with some high-speed ESCs that have been programmed with the so-called SimonK firmware. The SimonK firmware is named after Simon Kirby, who created a high-speed software that can be flash-programmed into most ATmega8L-controlled ESCs. Reprogramming flash memory in an ESC is done by using the six ISP pins, which are shown at the bottom left of the ESC PCBs in Figure 5.11. I would not recommend attempting to reprogram your ESCs unless you have successfully done it before. Otherwise, you will likely brick them, that is, make them inoperative unless they are restored to their original firmware. You can refer to the following website to learn more about high speed ESCs: wiki.openpilot.org/display/Doc/RapidESCs.

Battery Eliminator Circuit

The *battery eliminator circuit* (BEC) refers to the three-wire lead set extending from the ESC. Typically the wires are color coded as follows:

1. white = signal
2. red = power
3. black = ground

Other colors are also used, including brown, red, and orange, where brown is ground, red is power, and orange is signal. I have previously referred to the BEC cable as a servo connection, which is partly true. In a normal servo cable, white is still signal, black is still ground, but red is a power consumer not a power supplier, as is the case with the BEC cable. This is why it is called a battery eliminator, as the BEC normally plugs into an R/C receiver and supplies the power to that receiver. This eliminates the need for a separate receiver power supply, hence the name BEC. I have extracted a portion of the detailed ESC schematic, which is posted online at www.mhprofessional.com/quadcopter, as Figure 5.26.

You should be able to identify two 7805 regulator chips whose outputs are connected in parallel. This combined output is connected to the BEC red wire. The 7805s are linear regulators that have been manufactured for many years by many companies. Each one can

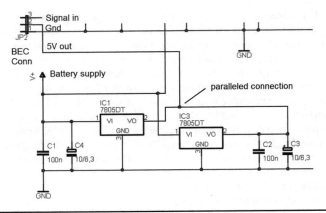

FIGURE 5.26   BEC schematic.

output a maximum current of 1 A at 5 V. There is a maximum of 2 A available at the BEC connector, since two regulators are paralleled. Two amperes is usually considered to be plenty of current to power an R/C receiver with several servos connected to it. Normally, there is not an issue when you connect a single BEC to an R/C receiver, which is the case for most R/C aircraft. An issue arises when four BECs are connected simultaneously to the same flight-control board. The flight-control board then uses the power provided by the ESCs to power the R/C receiver connected to it as well as any auxiliary servos that may be incorporated into the quadcopter. When you have the four BECs paralleled in a quadcopter, then eight 7805s are in turn paralleled. Some quadcopter designers think that having multiple BECs feed a common supply point can lead to trouble. This trouble could manifest itself as uneven power production whereby one BEC circuit would eventually take the whole load and overheat from the excessive current. It is my belief that there is no potential problem provided that all the ESCs are the same model and all use 7805s as regulators. After all, two 7805s are internally paralleled to increase the current capability, which leads me to believe that externally paralleling BECs will easily be tolerated, with each set of regulators providing a proportional amount of the load. The recommended solution for eliminating this potential problem is to cut all but one of the red BEC leads that are connected to the flight-control board.

CAUTION: Do not cut all the red BEC leads, or there will be no power flowing to the flight-control board.

I have operated my quadcopters both ways and encountered no problems with either one BEC powering the flight-control board or all four powering it. The only caveat arises if you use different model ESCs because they can have regulator circuits that are different from the very stable 7805-based units. If that is the case, I would recommend using only the single lead configuration.

Propellers

Propellers, while seemingly simple, are in reality, quite complex devices. They are airfoils that have been twisted to produce a thrust while rotating through an air mass. Propellers have been relatively unchanged since the days of the Wright brothers. Modern propellers are about 80% efficient, which is about the same as the Wrights experienced. Propeller efficiency is defined as:

$$H = \frac{propulsive\ power\ output}{shaft\ power\ input} = \frac{thrust + axial\ velocity}{resistance\ torque + angular\ velocity}$$

Basically, efficiency is about how much effective power is produced versus how much is input. Input, in our case, is the rotating outrunner motor, and output is the resultant thrust that is created. Efficiently operating propellers depend on an optimal *angle of attack* (AOA), which is mainly a function of how fast the propeller is rotating and how much power is being generated. In complex airplanes, the propeller AOA is adjustable; however, in our case, the propellers have a fixed pitch, or AOA. The AOA is set by the manufacturer in a compromise setting that tries to optimize efficiency for the expected operating range. There is really nothing you can do to improve propeller efficiency other than to try different models that have been designed to closely match your requirements. I think that the propellers provided in the Elev-8 kit are adequate to do what is needed. They are also relatively inexpensive, which is a key consideration, since you will need to replace them after flight mishaps.

Selecting a properly sized propeller is an important activity and a topic that has some degree of anecdotal qualities associated with it. Everyone has an opinion on this topic so I will offer some general guidelines:

- Thrust is proportional to the propeller surface area; therefore, larger propellers mean more thrust. When compared to the small-diameter propellers, the large ones also need more power in order to spin and achieve the same angular velocity.
- Conversely, small-diameter propellers need to spin faster to create the equivalent thrust of a large-diameter propeller.
- Quadcopters tend to hover more than to fly in straight lines. A smaller pitch, or AOA, is better suited for hovering than for aerobatic operations.
- Propellers should always be balanced to reduce vibrations. Investing in a quality propeller balancer is a good idea.
- Carbon-fiber propellers are stronger than and vibrate less than plastic propellers. Unfortunately, they cost quite a bit more than the plastic. Hold off on this investment until you improve your flying skills and lessen the amount of propeller damage.

Table 5.5 shows some standard propeller sizes used in quadcopters to aid you in selecting a propeller. It is very important to determine what functions you want your quadcopter to

Model	Diameter	Pitch	Description
APC 1047	10	4.7	Popular size for midsized quadcopters. This is the type used in the Elev-8
RPP 1045	10	4.5	Another popular size for midsized quadcopters
EPP 1245	12	4.5	Large propeller suitable for larger quadcopters
EPP 0938	9	3.8	Small propeller suitable for small quadcopters
EPP 0845	8	4.5	Small propeller often used on small quadcopters

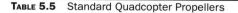

TABLE 5.5 Standard Quadcopter Propellers

perform before you select a motor and propeller combination. Choosing a video platform would lead you to select a propeller with a high pitch and/or large diameter so that it could turn at a lower r/min to produce enough lift for the camera, while minimizing vibrations. You should look at smaller diameter propellers with medium pitch if you are looking for a quadcopter that can perform aerobatics.

Comprehensive Quadcopter Analysis

I found an interesting website that will interactively create a set of performance characteristics for your quadcopter based on information you provide: http://www.ecalc.ch/xcoptercalc .htm?ecalc&lang=en. All you need to do is input these parameters:

- Overall quadcopter weight
- Battery type
- Select motor from a database list
- Select propeller from a database list

Then click on the Calculate button and you will see the results shown in Figures 5.27 and 5.28.

A wealth of information is in the calculation results, with much of it reflecting discussions presented in this chapter. However, I suggest that you use the website data with caution, since it has not been independently reviewed. However, it all appears to be reasonable and useful.

When reviewing the motor characteristics shown in Figure 5.28, I noticed that the motor achieves about 80% efficiency when operating at 3 A or above. I also saw that the motor

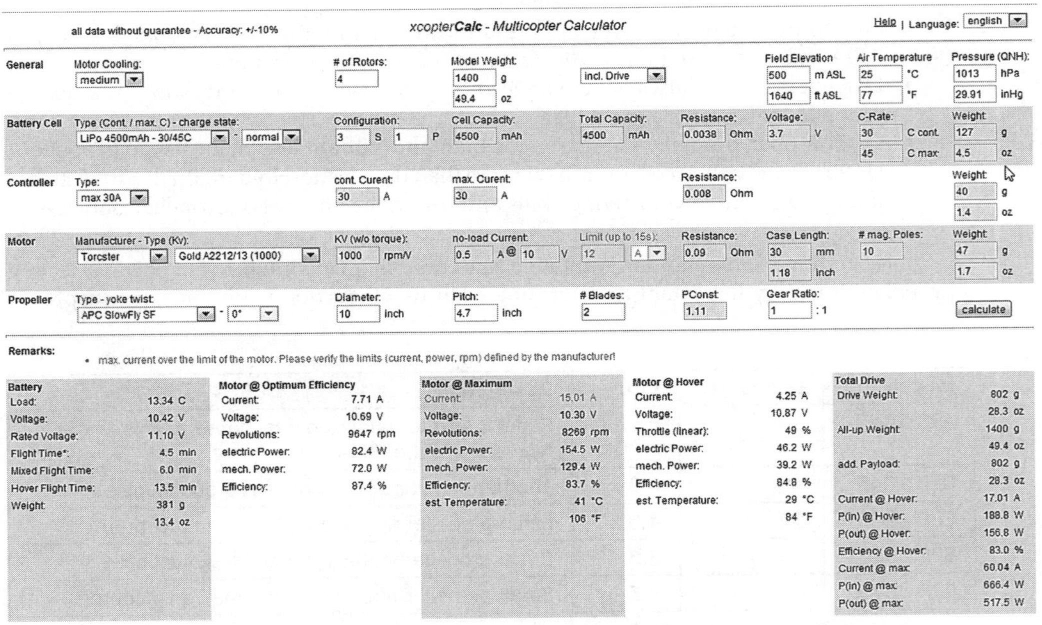

Figure 5.27 Calculation results from the interactive xcopterCalc website.

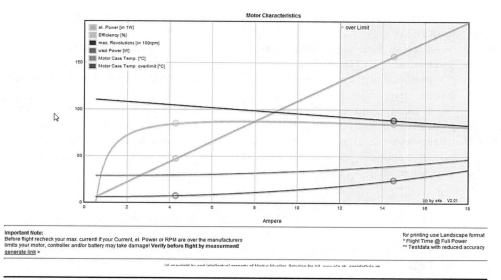

FIGURE 5.28   Characteristic motor graphs from the xcopterCalc website.

temperature is not predicted to rise too appreciably even when operating above the manufacturers recommendations. However, I am not convinced that it will remain cool if operating in the extreme regions.

This last section concludes my regular discussion on the quadcoptor propulsion components. What follows is a detailed program analysis of the ESC_Motor_Control_Demo program that was used in the experiment. It is presented for those readers who are obviously interested in such an analysis and who might want to use this program as a template for further experiments, and perhaps, a future flight-control program. The discussion also explores one Propeller Assembly Language (PASM) routine because it is used in the demo program. Finally, I delve into a brief C language discussion in which I demonstrate how a simple Propeller C Language routine can replace the somewhat obtuse PASM routine. Those readers not interested in such minutia can skip the following sections without fear of losing any continuity in the overall discussions.

ESC_Motor_Control_Demo Analysis

This analysis begins with the code listing. The program was made available on Parallax's Propeller forum, which is a very valuable resource of expert knowledge. Members of the forum can answer just about any of your questions regarding the Propeller chip, or even the Elev-8 for that matter.

```
''    Single_Servo_Assembly
''    Author: Gavin Garner
''    November 17, 2008
''    See end of file for original comments
'_____,

'   Modifications by Cluso99.
```

' Modified to drive a Brushless Motor using an ESC/BEC motor controller.
' Note a BEC type ESC supplies 5V on the center servo pin (some servos are different).
' Therefore, only two wires should be connected, being ground and servo pin.
" I have used the TV output because it has a series resistor for protection and ground outputs. I have connected an RCA plug to a servo type cable & plug from an old PC and a 3pin stake block. You could also put a series resistor in the RCA plug for protection if you wish."

" Here is the connection diagram...
     NOTE: *Deleted from this copy, see original if needed*"
'_____'

' Uses the PropPlug and PSerialT to set the motor speed from the PC Keyboard
'   0=OFF, 1=12.5%, 2=25%, 3=37.5%, 4=50%, 5=62.5%, 6=75%, 7=87.5%, 8=ON
'_____

" RR20100426   _rr001 use 20ms and 1ms...2ms (motor 1ms=off, 1.5ms=50%, 2ms=100%)"
" uses fdx and PSerialT: 0=OFF, 1=12.5%, 2=25%, 3=37.5%, 4=50%, 5=62.5%, 6=75%,7=87.5%, 8=ON"

```
CON
  _xinfreq  = 5_000_000
  _clkmode  = xtal1+pll16x      'The system clock is set at 80MHz
                                 (rec'd for optimal res)

  Servo_Pin = 14                ' use the TV pin (My note: servo
                                   pin 14)

  rxPin  = 31                   ' serial
  txPin  = 30
  baud   = 115200
  tvPin  = 14                   'TV pin (1-pin version)
  kdPin  = 26                   'Kbd pin (1-pin version)


OBJ
  fdx   :    "FullDuplexSerial"    'serial driver


VAR
  long  position     'The assembly program will read this variable
                      from the main Hub RAM to determine the servo
                      signal's high pulse duration

PUB Demo | ch
```

```
    waitcnt(clkfreq*5 + cnt)                    'delay (5 secs) to get
                                                 terminal program running
    fdx.start(rxPin,txPin,0,baud)               'start serial driver to PC
    fdx.str(string(13,"Cluso's Motor Control Test v002",13))
    fdx.str(string("Press <space> to start"))
    repeat
      ch := fdx.rx
    until ch := " "
    fdx.tx(13)
    position := 80_000                          '1ms (motor off)

    cognew(@SingleServo,@position)   'Start a new cog and run the
                                      assembly code starting at the "SingleServo"
                                      cell 0 and passing the address of the
                                      "position" variable to that cog's "par" reg
    repeat
      ch := fdx.rx
      case ch
        "0" : position := 80_000                     '1ms        OFF
        "1" : position := 90_000                     '1.125ms 12.5%
        "2" : position := 100_000                    '1.250ms 25%
        "3" : position := 110_000                    '1.375ms 37.5%
        "4" : position := 120_000                    '1.5ms    50%
        "5" : position := 130_000                    '1.625ms 62.5%
        "6" : position := 140_000                    '1.75ms   75%
        "7" : position := 150_000                    '1.875ms 87.5%
        "8" : position := 160_000                    '2ms         ON
```

```
DAT
'The assembly program below runs on a parallel cog and checks the
value of the "position" variable in the main hub RAM (which other
cogs can change at any time).
'It then outputs a servo high pulse for the "position" number of
system clock ticks and sends a 20ms low part of the pulse.
'It repeats this signal continuously and changes the width of the
high pulse as the "position" variable is changed by other cogs.

org        'Assembles the next command to the first cell (cell 0) in
            the new cog's RAM
SingleServo  mov      dira,ServoPin          'Set the direction of the
                                              "ServoPin" to be an output (and
                                              all others to be inputs)

Loop          rdlong   HighTime,par          'Read the "position"
                                              variable (at "par") from main
                                              RAM and store it as "HighTime"
```

```
        mov     counter,cnt        'Store the current system
                                    clock count in the
                                    "counter" cell's address

        mov     outa,AllOn         'Set all pins on this cog
                                    high (really only sets
                                    ServoPin high because
                                    all the rest are inputs)

        add     counter,HighTime   'Add "HighTime" value to
                                    "counter" value

        waitcnt counter,LowTime    'Wait until "cnt" matches
                                    "counter" then add a 20ms
                                    delay to "counter" value

        mov     outa,#0            'Set all pins on this cog
                                    low (really only sets
                                    ServoPin low because all
                                    the rest are inputs)

        waitcnt counter,0          'Wait until cnt matches
                                    counter (adds 0 to
                                    "counter" afterwards)

        jmp     #Loop              'Jump back up to the cell
                                    labled "Loop"

'Constants and Variables:
ServoPin    long    |< Servo_Pin   'This sets the pin that
                                    outputs the servo signal
                                    which is the white wire
                                    on most servomotors).

AllOn       long    $FFFFFFFF      'This will be used to set
                                    all of the pins high

LowTime     long    1_600_000      'This is a 20ms pause time
                                    for a 80MHz system clock.

counter     res                    'Reserve one long of cog
                                    RAM for this "counter"
                                    variable

HighTime    res                    'Reserve one long of cog
                                    RAM for this "HighTime"
                                    variable

            fit                    'Makes sure the preceding
                                    code fits within cells
                                    0-495 of the cog's RAM
```

{{Copyright (c) 2008 Gavin Garner, University of Virginia
Single_Servo_Assembly

Software is furnished to do so, subject to the following conditions: The above copyright notice and this permission notice shall be included in all copies or substantial portions of the Software. The software is provided as is, without warranty of any kind, express or implied, including but not limited to the warranties of non-infringement.

In no event shall the author or copyright holder be liable for any claim, damages or other liability, out of or in connection with the software or the use or other dealings in the software.

Notes

This program demonstrates how to control a single R/C servomotor by dedicating cog-to-output signal pulses, using a simple assembly program. Once the assembly program is loaded into a new cog, it continuously checks the value of the "position" variable in the main RAM (the value of which can be changed at any time by code running on any other cog) and creates a steady stream of signal pulses with a high part that is equal to the value of the "position" variable times the clock period (1/80 MHz) in length and a low part that is 10 ms in length. (This low part may need to be changed to 20 ms depending on the brand of motor being used, but 10 ms seems to work fine for Parallax/Futaba Standard Servos and gives a quicker response time than 20 ms.) With an 80-MHz system clock, the servo-signal's pulse resolution is between 12.5–50 ns; however, the control circuitry inside most analog servomotors probably will not be able to distinguish between such small changes in the signal.

To use the above code in your own Spin code, simply declare a "position" variable as a long, start the assembly code running in a cog with the "cognew(@SingleServo,@position)" line, and copy and paste my DAT section into the DAT section of your own code. Note that you must change the number "7" in the ServoPin constant declaration in the assembly code to select a pin other than Pin 7 to be the output pin for the servo signal.

If you are using a Parallax/Futaba Standard Servo, the range of signal-pulse widths is typically between 0.5–2.25 ms, which corresponds to "position" values between 40_000 (full clockwise) and 180_000 (full counterclockwise). In theory, this provides you with 140_000 units of "position" resolution across the full range of motion. You may need to experiment with changing the "position" values a little to take advantage of the full range of motion for the specific R/C servo motor that you are using. However, you must be careful not to force the servo to try to move beyond its mechanical stops. If you find that your propeller chip or servomotor stops working for no apparent reason, it could be that the motor is sending inductive spikes back into the power supply or it is simply drawing too much current and resetting the propeller chip. Adding a large capacitor (e.g.,1000 uF) across the power leads of the servo motor, or using separate power sources for the propeller chip's 3.3-V regulator and the servomotor's power supply will help to fix this.

The Spin portion of the program uses the FullDuplexSerial object that sets up a means by which the user can input data via the keyboard as well as view data on the PC screen. The PSerT program must first be running on the PC, as was discussed above, and then the FullDuplexSerial object will take over all the duties of communicating between the program and the PC. The FullDuplexSerial object has the local reference name of fdx, which is short for full duplex. The fdx object is preset for a 115,200 baud rate as well as using the normally designated Prop chip pins 31 for receive and 32 for transmit. These pin designations are standard on all Parallax Propeller development boards and match the Prop chip's functions.

I would like to point out the presence of a "magic number" in the Spin statement that starts the fdx object:

```
fdx.start(rxPin,txPin,0,baud)          'start serial driver to PC
```

I found the meaning of this magic number in the website http://propeller.wikispaces .com/Full+Duplex+Serial. The 0 in the argument list represents the operating mode for the full duplex serial object. Below is the wiki documentation regarding the mode:

```
.start(rxPin, txPin, mode, baudrate)
```

> *Start serial driver—starts a cog*
> *mode bit 0 = invert rx*
> *mode bit 1 = invert tx*
> *mode bit 2 = open-drain/source tx*
> *mode bit 3 = ignore tx echo on rx*

I guess it is just the teacher in me, but I would prefer to eliminate all magic numbers and instead modify this code to include a constant definition as follows:

In the CON section add

```
mode = 0
```

and in the PUB section modify

```
fdx.start(rxPin, txPin, mode, baud)          'start serial driver to PC
```

You see how much clearer it all becomes, although it is true that you will have to go back into the documentation to discover exactly how the mode works. Please refer to my rant in the previous chapter regarding magic numbers. Now back to the analysis.

There is a "repeat – until" loop checking for a character from the PC. This code snippet is shown below:

```
repeat
  ch := fdx.rx
until ch := " "
```

All you need to do is press the space bar in order to enable you to enter the desired power level. The motor always starts at the 0 power level meaning, no rotation. The Spin program will then launch a new cog after the initial position value (80,000 in this case) is stored. The following code snippet creates this new cog object:

```
cognew(@SingleServo,@position)
```

This statement creates a new cog that runs code beginning at the symbolic address "SingleServo," which is also the start of an assembly language program. The statement also instructs the cog to load, at boot time, the value stored in the hub RAM location named "position" into its "PAR" *special purpose register* (SPR). I introduced the SPRs at the end of Chapter 4, but I did not really elaborate on them. The PAR SPR is the so-called boot register, which means that a value will be stored in it when the cog is created and/or rebooted when it is executing assembly code. You should note that creating a cog that is designated to execute assembly language is substantially different from creating a cog to run a method in another Spin object, which was the situation in all of the Chapter 4 examples. The

"@SingleServo" is the way Spin knows to go to a specific memory location in a cog's RAM to load and execute instructions. By default, the cog memory location is normally set at 0, which is explained below in the assembly language discussion.

An infinite loop then runs and checks for a new character appearing in the fdx buffer. It will subsequently store the clock-cycle count equivalent to the pulse width desired in the "position" variable. Remember from the above experiment that 1 ms is 0 or no rotation, while 2 ms is 100% or maximum rotation. 1 ms is equal to 80,000 clock cycles, while 2 ms is equal to 160,000 clock cycles, all with respect to an 80-MHz system clock. The specific cycle number is stored in a designated memory location in the hub RAM memory named "position," as previously discussed. Thus, the commanded pulse width is always available to the newly created cog using the PAR SPR. Also, recall that the cogs are constantly being refreshed by the hub; thus, any new values appearing in a shared hub memory location will almost instantly appear in the appropriate cog SPR.

Next is a discussion of the few assembly language instructions that are part of this demo program. I have repeated them below so you do not have to constantly refer to the original code. I also deliberately removed all the comments so you could focus on the actual instructions as well as the numbering of the lines for an easier reference. I will go through the instructions one at a time in Table 5.6, since that is probably the least confusing way to approach this somewhat complex subject.

```
1 org
2 SingleServo       mov       dira, ServoPin
3 Loop              rdlong    HighTime, par
4                   mov       counter, cnt
5                   mov       outa, AllOn
6                   add       counter, HighTime
7                   waitcnt   counter, LowTime
8                   mov       outa, #0
9                   waitcnt   counter, 0
10                  jmp       #Loop
11 ServoPin         long      |< Servo_Pin
12 AllOn            long      $FFFFFFFF
13 LowTime          long      1_600_000
14 counter          res
15 HighTime         res
16                  fit
```

The remaining line item not included in Table 5.6 is line 16, fit, which is not an instruction per se, but an assembler directive just like org is in line 1. The fit directive ensures that all the data and instructions are nicely aligned as four byte packages or words, since that is how the cog memory must be configured.

The last remaining line item to be further discussed is line 11:

```
ServoPin        long      |< Servo_Pin
```

The value Servo_Pin is defined in the Spin program CON section as the decimal value 14. This value must be decoded into a specific bit position, which is the purpose of the bitwise

Line Number	Instruction	Explanation
1	org	An assembler directive that establishes the starting point in cog memory where all the succeeding instructions and data will be stored. Normally 0 as it is in this case
2	SingleServo mov dira, ServoPin	The identifier, SingleServo, is assigned the first memory address which is 0. The mov instruction takes the source data value, ServoPin, and copies it into the cog's GPIO data direction register, dira.
3	Loop rdlong HighTime, par	The identifier, Loop, is assigned the next available address, which is also the start of a loop. The rdlong instruction copies the long value from the par register into the HighTime variable that was created by line 15.
4	mov counter, cnt	The system counter value, cnt, is copied into the counter variable created by line 14.
5	mov outa, AllOn	The constant AllOn setup in line 12 is copied into the cog's GPIO data register, outa. This instruction turns on pin 14.
6	add counter, HighTime	The value stored in the variable, HighTime, is added in two's complement fashion to the value in the variable, counter.
7	waitcnt counter, LowTime	Wait until the system counter, cnt, matches the current value in counter, and then add the value of LowTime (defined in line 13) to the counter value
8	mov outa, #0	Move the immediate value 0 to the cog's GPIO data register, outa. This will cause a zero or low output on pin 14.
9	waitcnt counter, 0	Wait until the system counter, cnt, matches the current value in counter and then add the value of 0 to the counter value. This effectively keeps the pin output low for the LowTime value, which is 20 ms.
10	jmp #Loop	Unconditionally jump back to the memory location named Loop that was defined in line 3.

TABLE 5.6   Assembly Language Routine Analysis

decode instruction " | <." The above bit operation results in the following binary bit pattern being stored into the ServoPin variable:

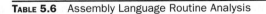

%00000000 00000000 01000000 00000000

This bit pattern is copied into the cog's dira register, thereby enabling pin 14 as an output, which is the desired outcome.

A Brief Introduction to the C Language

You should probably appreciate a simpler alternative to creating fast code after going through the above assembly language discussion. Well, you are in luck, as Parallax has recently made available a C language development environment for the Prop chip. The C language has been around for many years, originating in 1969 thru 1973. It has evolved and transformed to the point where it now supports many different types of computing platforms ranging from high-end servers to the minimalist embedded microprocessor. The Prop C development software is named SimpleIDE and is available free of charge from the Parallax "learn" website http://learn.parallax.com/.

It is a fairly simple download-extract-and-install process. My only caution is that you must be sure that the compiler, library, and workspace locations are properly identified. You do this by clicking on Tools in the SimpleIDE menu bar and then on the Properties selection. A dialog box, as shown in Figure 5.29, will appear. You should see locations similar to the one in this figure if you installed the IDE using the default settings.

I elected to load one of the example programs to test out my SimpleIDE installation. Feeling somewhat confident, I loaded the program named Standard Servo Position.c that is included in an Examples library that may be downloaded from the Parallax "learn" website. My computer's location for the directory holding this program was: c:/Program Files (x86)/SimpleIDE/Workspace/Learn/Examples/Devices/Motor/Servo.

I included the above path only to indicate that it is relatively easy to lose track of where all your project, source, library, and include files are located. Losing track of the location of these files may easily be a source of great frustration when you start getting compile-time errors that a given file could not be found. I suggest you follow all the good guidelines provided in the Parallax "learn" website http://learn.parallax.com/ regarding how to setup the SimpleIDE.

I guess I was a bit lucky in that it took only two attempts to successfully compile and download the example servo-control program into the BOE. Figure 5.30 is a screenshot of the SimpleIDE with the source code as well as most of the compiler and loader reports, which are shown below the source code editor pane.

The example program is very simple in that it directs a standard Hitec servo to move to several positions and then pause for three seconds at each position. The servo was connected

Figure 5.29 SimpleIDE properties dialog.

File Project Edit Tools Program Help

```
Standard Servo Position.c *
 9  #include "simpletools.h"
10  #include "simpletext.h"
11  #include "simplei2c.h"
12  #include "servo.h"
13  #include "simpletools.h"                    // Include simpletools header
14  #include "servo.h"                          // Include servo header
15
16  int main()                                  // main function
17  {
18      servo_angle(16, 0);                     // P16 servo to 0 degrees
19      pause(3000);                            // ...for 3 seconds
20      servo_angle(16, 900);                   // P16 servo to 90 degrees
21      pause(3000);                            // ...for 3 seconds
22      servo_angle(16, 1800);                  // P16 servo to 180 degrees
23      pause(3000);                            // ...for 3 seconds
24      servo_angle(16, 450);
25      pause(3000);
26      servo_angle(16, 0);
27      pause(3000);
28      servo_stop();                           // Stop servo process
29  }
```

Figure 5.30 Screenshot of the SimpleIDE with the example servo program loaded.

to the BOE P16 servo port with the source code reflecting that selection. Figure 5.31 shows the BOE and servo connected.

You should also understand that the SimpleIDE represents a substantial development effort by Parallax engineers as well as many others who develop and maintain the open-

Figure 5.31 BOE and standard servo setup for the example program.

source GCC compiler that the SimpleIDE uses to compile and load the source code. I have used GCC for many years but never with as much ease as I experienced with the SimpleIDE. Well done Parallax!

Summary

First, congratulations as you have slogged through this somewhat complex but (I hope) interesting chapter. I believe you will have gained a substantial appreciation of the Elev-8 propulsor components that should help you understand and evaluate new choices when (not if) it comes time to modify your quadcopter.

The chapter began with a detailed examination of the highly energetic and somewhat unusual motors that power quadcopters. These motors are known as outrunners because the rotors are on the outside of the motor, while the stator is stationary on the inside (a bit unusual as compared to normal electrical motors). These motors are also multiphasic—they are driven by dedicated controllers known as *electronic speed controllers* or ESCs. The motors are very light and compact but can produce extreme amounts of power for their size, which is why quadcopters can fly.

The ESC that powers the motor was discussed next with attention drawn to the ATmega8L microcontroller that controls the ESC. I also explained how the raw battery power is switched on by a series of power MOSFETs to provide the three-phase power for the motor. ESC waveforms were shown to help illustrate how the three-phase power technique works.

A lengthy discussion followed regarding an experiment that I designed to show how an Elev-8 motor functioned with one of the propellers mounted on it. I went through the setup of the experiment and explained the control circuitry that was centered on the Parallax Propeller *Board of Education* (BOE) as the controlling system. The program running in the BOE was also thoroughly examined later in the chapter.

I discussed all the experiment's results and used a series of charts to help explain what caused certain outcomes and why. The results also provided some useful information regarding power consumed versus available flight time and other operational tradeoffs that should always be considered

Next came a brief discussion on ESC update rates and how some ESC designers are concerned about slow updates.

I discussed the *battery eliminator circuit* (BEC), describing its design and purpose. I also pointed out some potential issues with BECs that are applicable only in a quadcopter design. A solution was also offered for those readers so inclined to follow it.

A section on the design and selection of propellers was next. Here I included a handy table of standard propellers that are commonly used in quadcopter designs. A set of propeller selection guidelines was also provided to help you choose knowledgeably.

The propulsor discussion ended with a brief introduction to a comprehensive and interactive website that allows you to conduct a detailed analysis on your quadcopter design. It is an incredible tool that should be used with care due to the great amount of information and data presented.

The remaining section of the chapter concerned a detailed analysis and discussion of the BOE control program that ran the experiment. I discussed the program in two major divisions, the first being the Spin code and the second, the assembly language code. These discussions were initiated to further increase your background in the Propeller languages that began in Chapter 4.

I recognized that the assembly language discussion would be a bit daunting to readers who are not too conversant in imperative programming. I still wanted to introduce the topic and take you through it as gently as possible. The good news is that now an alternative to assembly language exists, and it is the C Language for the Propeller.

The chapter concludes with a very brief and concise introduction to the C language that has been adapted or "ported" to run on a Prop chip. It runs in an environment known as the SimpleIDE, and it is very easy to use. Take my word on this last statement, as I have used C tool chains (environments) for the last 30 years, and this one is definitely the best, at least for the embedded market. I showed a reasonably complex application that controlled a standard servo using the BOE. I had this application up and running within 30 minutes, which is almost instantaneously in software development time.

The next chapter takes an interesting view of what makes up a modern radio-controlled (R/C) system and how it functions. There are some important points that you should be aware of when operating an R/C system in today's crowded spectrum. You don't want to lose control of your quadcopter, and unfortunately, it is easier to do than you think. So read on.

CHAPTER 6

Radio-Controlled Systems and Telemetry

Introduction

The remote control by radio waves of devices ranging from battleships to insect-sized flying machines has been ongoing since Nikola Tesla's experiment in 1898. I will not repeat the history here but instead refer you to the interesting Wikipedia entry at http://en.wikipedia .org/wiki/Radio_control. This chapter focuses on how a modern 2.4-GHz *radio-controlled* (R/C) system functions and explores additional features that will make your quadcopter flying experience more enjoyable, and maybe even a little educational.

Evolution of Model R/C Systems

The R/C systems used to control model aircraft first appeared in the early 1950s. The practical reason for this was the advent and availability of cheap transistors. Up until that time period, radio systems were built using vacuum tubes that required bulky components and batteries, neither of which could be easily placed in small model aircraft. Transistor circuits changed all of that because they require only a little battery power, run cool, and take up very little space. Over time, discrete transistors gave way to integrated circuits, which eventually morphed into the powerful microcontrollers that are in practically all modern R/C transmitters and receivers. Of course, the servos, which are the mechanical actuators being controlled by the R/C receiver, also changed from relatively large and heavy units to very lightweight yet powerful units. I will discuss servos in the next chapter, since there is plenty of material to cover regarding their use and function.

The best way to understand the modern R/C system is to start with a discussion about the basics that underpin any radio system. I am not going to make this a tedious tutorial but will try to hit the high points to provide you with a reasonable idea of what makes the 2.4-GHz system tick.

Carriers and Modulation

All radio communications use a wave known as a carrier. This is the fundamental electromagnetic wave that is created at the transmitter to carry information or data to a compatible receiver. Carrier waves normally do not have any information impressed on

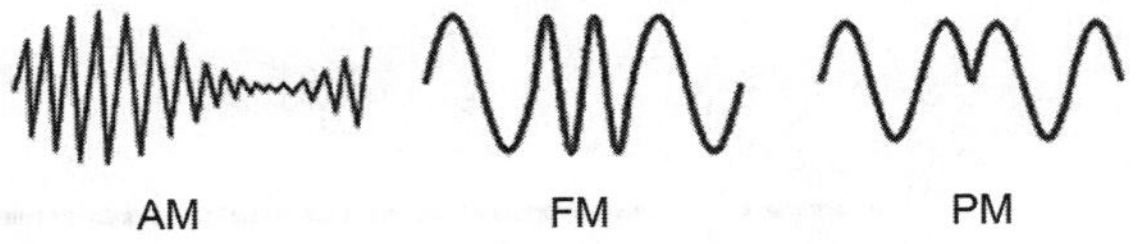

FIGURE 6.1 AM, FM, and PM modulation waveforms.

them but must be modulated or altered in some standard fashion in order to send data. There are three principal modulation methods:

1. *Amplitude modulation* (AM)
2. *Frequency modulation* (FM)
3. *Phase modulation* (PM)

Figure 6.1 shows the waveform representations for these three modulation methods.

Many more modulation schemes exist today, but they are all dependent on some combination of AM, FM, or PM. Most of the relevant modulation schemes that currently use the R/C field are shown in Table 6.1. Sharp-eyed readers might be wondering why I didn't include *pulse-width modulation* (PWM) in the list in Table 6.1. After all, I did discuss it in great detail in the previous chapter. The answer is that PWM is actually handled as part of *pulse-position modulation* (PPM). Figure 6.2 should help clarify how this is accomplished.

Typical pulses used for R/C servo-control signals have pulse widths that vary from 1 to 2 ms and repeat every 20 ms, as shown on the left side of Figure 6.2. That means as much as 18 ms of time is wasted, or not utilized, if only one pulse is sent per 20-ms cycle, or frame as it is called in PPM terminology. PPM overcomes this limitation by sending all of the servo channel pulses, one after the other with no wasted space between them, as shown on the

Type	Abbreviation	Name	Description
Analog	AM	Amplitude Modulation	Changes the carrier wave amplitude proportional to the data.
Analog	FM	Frequency Modulation	Slightly changes the carrier frequency proportional to the data.
Digital	PPM	Pulse-Position Modulation	Changes the pulse position in a frame proportional to the data.
Digital	PCM	Pulse-Coded Modulation	Sends digital data describing the data.
Spread Spectrum	DSSS	Direct-Sequence Spread Spectrum	Sends PCM data over a spectrum range with error corrections.
Spread Spectrum	FHSS	Frequency-Hopping Spread Spectrum	Sends PCM data using synchronized carrier frequencies that hop throughout a spectrum using a pseudorandom sequence.

TABLE 6.1 Common R/C Modulation Techniques

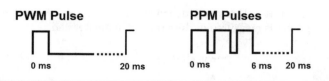

FIGURE 6.2   PWM and PPM.

right side of Figure 6.2. PPM efficiently utilizes the radio-frequency spectrum with up to 10 channels available in a 20-ms frame in which each channel occupies only a 2-ms slot. I am sure that even the total frame period is adjusted to accommodate the maximum number of channels being sent, since again, it makes no sense to waste 10 ms every frame if there are only 5 channels available to be transmitted.

After receiving the PPM stream, the R/C receiver restores each servo channel and ensures that each channel has only the required one pulse for every 20-ms period. Modern R/C receivers usually use a microprocessor to control this process.

PCM is quite a bit different from PPM, since each channel's data is sent using data bits that represent the value of the control function. The actual value is dependent upon the number of bits used to encode the position of the transmitter control being used. Let's consider a throttle-stick control in which 10 bits may typically be used to encode the relative position of the throttle. There are 1024 positive integer numbers that can be represented by 10 bits, which is equivalent to 2^{10} power. Therefore, 0 would be the 0% throttle position, while 1023 would be the 100% throttle position. Sending this number via PCM to the receiver ensures an extremely accurate representation of the equivalent throttle-control position, as it is set on the transmitter. PCM is the quality method that all reliable R/C systems use.

Not all transmitter controls need this precision. Consider the gear control where only two values are required, say 0 for gear up and 1 for gear down. It is nonsensical to send 10 bits for this function where 1 bit would suffice. PCM systems are optimized to account for the transmission of varying channel data precision where some channels require many data bits for the encoded channel data, while others require only a few.

Noise

Noise, or electrical interference, is the single most important problem with R/C systems. Noise present in an operating area could easily cause you to lose control of your aircraft. This could result in you losing the aircraft, or, even worse, causing injury to nearby spectators and/or damage to property. You would be liable for injury and damages even though you were not at fault for the interference that caused the loss of control. You always want to avoid these unfortunate scenarios, which is the reason that some very robust and secure modulation schemes were developed. Before I discuss these schemes I want to show you how noise does affect some of the basic modulation techniques.

One item that is common to all types of modulation is the loss of signal. You really cannot ascribe this to interference; instead, the root cause is simply having a signal strength that is presented too low at the receiver antenna. A low signal may be caused by several factors, including operating the aircraft too far from the transmitter. The loss of signal strength is due to spherical spreading, in other words, the further you are from the transmitter the more diluted or weakened the signal becomes. Remember this spherical-spreading rule of thumb (sometimes called the square law): twice the distance from the transmitter means one quarter of the original signal strength at the receiver. For example, if you had one unit of strength at a distance of 1 meter from the transmitter, there would be only 1/16,384 of the original strength at a 128 meter distance. That is a considerable reduction. Transmitter strength or power is measured in dBm units that are defined as follows:

$$\text{power dBm} = 10 \log (\text{power in milliwatts})$$

Therefore, 1 milliwatt, or 1 mW, is equivalent to 0 dBm. Strictly speaking, the impedance used in the dBm measurement should be 300 Ω resistive, but if it is not, the measurement is still largely valid.

The Spektrum DX-8 transmitter that is used in my Elev-8 system has a maximum contact output (almost touching the antenna) of −10 dBm. That is not a lot of absolute power, but it is sufficient to its task. At one meter, the actual measured power is −25 dBm for the DX-8. Every doubling of the distance per the spreading rule of thumb means a linear reduction of 10 dBm, which is why using dBm units is so handy. If you use the example from above, the 128 meter distance would mean that the transmitter power at the receiver antenna would be −10 + (−70), or an absolute −80 dBm. This is a very small power level but remarkably well within the Spektrum AR8000 receiver's capability.

The other major culprit in signal loss is the line-of-sight restraint. The 2.4-GHz signals operate in the RF frequency region where signals travel in a straight line or *line of sight*. If you cannot see your aircraft, you can be pretty sure that it will not be able to receive your signal. A simple maneuver, such as flying your aircraft to the side of your house, will likely cause loss of signal due to losing line of sight. However, although it is true that the signal could reflect off a nearby object, such as a neighbor's house, I would not bet my quadcopter on that happening. Obviously, flying your quadcopter so far away that you can no longer see it would be bad on several levels, as discussed previously. You would also be flying it too high and breaking some civil regulations to boot.

AM and FM Modulation and Noise

AM is probably the one modulation type that is most susceptible to noise because almost any nearby electrical source can generate spurious radio waves. You are probably familiar with the somewhat noisy AM radio in your car. It will often pick up noise from your car engine as well as from nearby cars. The same condition may happen if you are operating your model outside your home. For instance, a nearby gas lawn mower could generate noise across a broad spectrum, especially if the spark plug is unshielded. AM receivers have no means for detecting and counteracting interfering signals that may cause the devices they are controlling to go out of control. AM is often used in extremely low-cost R/C toys, which is fine, since they are usually quite small and will not harm people or things if they suddenly go out of control. However, AM-based systems are to be avoided in the Elev-8 system or any other quadcopter application. Using an AM system is to invite disastrous consequences.

FM is much more resistant to noise, as you might realize if you think about the car radio example again. FM radio stations always sound clearer and mostly noise free when compared with AM stations. Part of this clarity has to do with the broader spectrum allocated to an FM station as well as to the nature of the modulation process. FM R/C systems use what is known as a *narrow band* (NB) channel in which the carrier frequency is slightly changed in response to the data-amplitude change. The narrower the frequency change or deviation, the more susceptible the FM channel is to interference. You should understand that the interference results from the receiver losing its lock, or phasing relationship, with the carrier wave due to noise, not to the introduction of pops or clicks in the signal. Interference is most often due to multiple reflections of the signal from buildings and terrain.

Another common source for FM interference is the presence of other NB FM transmitters operating on the same frequency. It is commonly referred to as cross-channel interference.

An R/C transmitter has no way of discriminating which signal it should respond to, so it will attempt to respond to them all, if possible. What happens, as you may imagine, is chaos where uncommanded actions may happen or the receiver simply does not respond at all. There is really no easy way to counteract cross-channel interference other than by shutting down all the interfering transmitters or relocating to a quiet location where you are operating the only transmitter.

The nature of PM interference is almost identical to that of FM interference, as both are quite similar in how they respond to interference.

Direct-Sequence Spread Spectrum

Direct-Sequence Spread Spectrum (DSSS) is the modulation technique used in the Spektrum DX-8 transmitter in my Elev-8 project. Before I start my discussion, I would like to point out that Spektrum calls their modulation technique DSM2, which is simply a marketing term for the DSSS standard. As far as I can determine, DSM2 is not different from DSSS; however, Spektrum is not at all forthcoming regarding any additions they may have added to the DSSS standard protocol. Therefore, I am assuming they are one and the same.

Further proof that DSM2 is DSSS is revealed by the transmitter module used in the DX-8, which may be seen at the top, center of Figure 6.3. The shiny silver case contains the Cypress Semiconductor CYRF6936 Wireless USB transceiver chip. The Cypress chip is fully compliant with the DSSS standard, which means that the Spektrum DSM2 must also be fully compliant.

FIGURE 6.3   Interior of Spektrum DX-8 transmitter.

Specification	Description
Frequency	2400 to 2483.5 MHz divided into eighty 1 MHz channels (the extra bandwidth is allotted to edge guard bands)
Maximum power	1000 mW US, 100 mW Europe, and 10 mW/MHz Japan
Minimum power	1 mW
Rx sensitivity	−80 dB

Table 6.2   Some Key R/C DSSS Specifications

DSSS is also the same modulation technique specified for use in the IEEE standard 802.11, commonly known as Wi-Fi. Some key R/C DSSS specifications are shown in Table 6.2.

As DSSS is fairly complex, I will attempt to describe and discuss only the essential features that are applicable to the R/C field. The essence of DSSS is to represent primary data symbols with another set of symbols that are spread out in time. Figure 6.4 is my rough sketch for this process.

The obvious question is why anyone would want to transform one symbol into many, as is shown in the Figure 6.4. The answer lies in circumventing the problems that arise when sending the primary symbols. Sending the primary, or raw, data is subject to noise and interference, and there is no means to detect and correct errors that happen during the transmission process. DSSS deliberately adds complexity to enable error detection and correction and to reduce the likelihood of noise corruption of the primary data. Transmitting the additional symbols also occupies more spectrum than just sending the primary symbols, as shown in Figure 6.5.

You may clearly see from Figure 6.5 that the primary data spectrum is highly concentrated around a specific frequency, while the DSSS is uniformly spread throughout the available spectrum, hence the name "spread spectrum." The tightly grouped primary data spectrum is more susceptible to noise than the spread spectrum.

Five processes are used in DSSS to minimize interference and ensure that only the data is sent and received between paired or bound transmitters and receivers. These processes are:

1. Automatic selection of dual transmit channels
2. Switching channels for every data frame transmitted

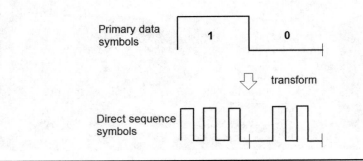

Figure 6.4   Primary data symbols transformed to direct-sequence symbols.

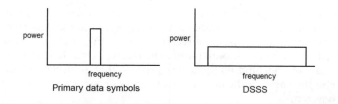

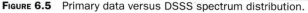

FIGURE 6.5 Primary data versus DSSS spectrum distribution.

3. Transmission of *start of packet* (SOP) and data *pseudorandom noise* (PN) packets
4. Transmission of two sets of *cyclic redundancy checks* (CRC)
5. Transmission of the *globally unique identifier* (GUID)

I will briefly examine each process to provide you with some insights on how DSSS functions.

Automatic Selection of Dual Transmit Channels

When the DX-8 (or any other DSSS compliant transmitter) first transmits, it selects a pair of channels from the 80 available. Theoretically, this would allow up to 40 DX-8's to be operated simultaneously in a small area without causing interference to each other.

Switching Channels for Every Data Frame Transmitted

The DX-8 will switch between the two channels for every data frame transmitted. This is a special form of frequency hopping that will be discussed in the FHSS section. Every data packet is transmitted twice, once on the first channel, then again on the second channel. It takes only several microseconds to switch channels.

Transmission of SOP and PN Packets

The DX-8 system uses a set of five 72-byte PN codes. A PN code that is prepended to the start of a data frame is called an SOP PN. A PN code that is prepended to the beginning of the "real" data packet is also called a DATA packet. These pseudorandom-coded data sets are so named because of how they are generated. True random data is exactly that: totally randomized so that the next data character cannot be predicted from the previous. Pseudorandom data, on the other hand, appears random in nature but is created by a predefined algorithm in which all the random codes are precisely generated in a deterministic fashion.

The receiver uses PN codes to determine whether or not to accept a particular transmission. When a PN code is unmatched, it means that a particular data package was not designated for that receiver and will be ignored. The SOP is 8 bytes long, while the DATA packet is 16 bytes long. Both the SOP and DATA PN packets are extracted from one of the five 72-byte PN sequences. Unique SOP/DATA pairs are also evenly distributed throughout all 80 channels to further lessen any potential interference between DX-8 transmitters.

Transmission of Two Sets of Cyclic Redundancy Checks

A *cyclic redundancy check* (CRC) is a calculated number based on both the numerical values contained in the whole data frame and a special manufacturing code value contained in the transmitter chip firmware (which is explained below in the GUID section). The CRC uses an error-correcting algorithm to create a two-byte number, using the frame data and the embedded code value. This number is then appended to the data frame and subsequently transmitted. The receiver uses the received CRC value and compares it to a recalculated value based on the received data frame. The receiver already knows the special embedded code because of the binding (pairing) process that is described in another section. A mismatch in the values indicates that a transmission error has occurred and the data frame must be rejected.

A second CRC is also created by exclusive ORing the data frame containing the first CRC with the hexadecimal value 0xFFFF. This just adds an additional error-checking capability to DSSS for further redundancy.

Transmission of the GUID

The *globally unique identifier* (GUID) is a two-byte value (for DSSS purposes) that is generated from a manufacturing code contained within the transmitter-chip firmware. The GUID for the DX-8 is based on the very unique manufacturing code created when the Cypress CYRF6936 chip was produced. This is very similar to how network adapter cards create *media access codes* (MAC) that uniquely identify a computer to the network to which it is attached. The MAC value is essentially the GUID for a networked computer.

The transmitter GUID is loaded into the R/C receiver during the binding process, which is why a DSSS transmitter-receiver pair will not function without doing this binding process. In addition, at least for the DX-8, all the positions and settings of the transmitter's controls at binding time are also stored in the receiver's memory. These are the fail-safe positions that will automatically be selected if the receiver loses connection with the transmitter.

The five DSSS processes practically guarantee that interference is eliminated and only the paired transmitter will function with its receiver. This is a big confidence booster that has promoted the DSSS standard among R/C enthusiasts. All currently available 2.4-GHz systems are extremely reliable because they use either DSSS or FHSS. The latter technology is discussed next.

Frequency-Hopping Spread Spectrum

Frequency-hopping spread spectrum (FHSS) adheres closest to the original *spread spectrum* (SS) concept that was invented and patented during WWII. Actress and inventor Hedy Lamarr, shown in Figure 6.6, originated the concept to help the Allies war effort.

Her patent envisioned the remote control of a torpedo with a radio carrier wave that hopped or skipped over 88 frequencies, which incidentally is the number of keys on a standard piano. She thought that the enemy would not be able to easily intercept or jam radio-controlled signals that were hopping about the spectrum. She was absolutely correct in her reasoning. It turned out that the U.S. government was not really interested in her invention and never adopted it for use in the war. Years later, it was widely adopted when researchers realized how robust SS was in minimizing corruption of communications from interception and interference.

A transmitter-receiver system using FHSS needs to be bound or paired in the same manner as a DSSS system. Most readers will be familiar with the *Bluetooth* (BT), which is designed to be a close-range *personal area network* (PAN). BT uses the FHSS modulation scheme to minimize interference, since many BT-equipped devices are often used in close proximity to each other. Of course, the power levels are much higher when FHSS is used for R/C purposes than when it is used for BT to couple your cell phone with a remote microphone/earpiece.

Figure 6.6   Hedy Lamarr, creator of the first spread spectrum patent.

Binding or Pairing

Binding for DSSS refers to the process in which the transmitter GUID and the fail-safe data are loaded into receiver firmware. The GUID is a parameter that uniquely identifies the transmitter to the receiver, and it is a prerequisite that the binding process be performed before any R/C operations take place.

Futaba is another major manufacturer of 2.4-GHz R/C systems that use FHSS as their modulation scheme. They term their FHSS technology as FASST, which stands for *Futaba Advanced Spread Spectrum Technology*. Their binding or pairing process consists of transferring the GUID and frequency hopping pattern from the transmitter to the receiver. Most readers are aware of the pairing operation that is normally required prior to using a non-R/C BT device. This BT pairing operation usually requires that the receiving device be put into a scan mode to identify any nearby BT transmitters. The users then enter a predefined code once the transmitter is identified and selected. The BT receiver will then load the transmitter's GUID and hopping pattern into the receiver's EEPROM so that it will no longer need to be paired the next time it is used with that particular transmitter. The actual Futaba binding process is very simple:

- Turn on the transmitter. Check the LED on the back of the transmitter to make sure that it is green. If so, proceed to the next step. If not, power down the transmitter and turn it on once again.
- Turn on the receiver.
- With the receiver on, press and hold the ID Set button (located between the two antenna exits) for more than one second. When the linking process has completed the receiver's LED will change to a solid green.

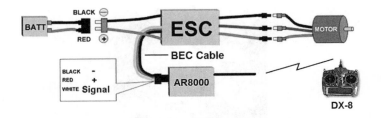

FIGURE 6.7   Diagram of the experimental R/C test system.

The procedure for binding a DX-8 to a Spektrum AR8000 receiver is a bit different from the BT pairing. I will use an experimental setup that is shown in diagram form in Figure 6.7 to demonstrate the process. The complete experiment will be discussed later in this chapter. The binding steps are listed below:

- Ensure that the battery is not connected to the ESC. The DX-8 should also be turned off.
- Put the bind plug into the BIND/DAT port on the AR8000. Figure 6.8 shows the plug in place and the ESC BEC plugged into the throttle port.

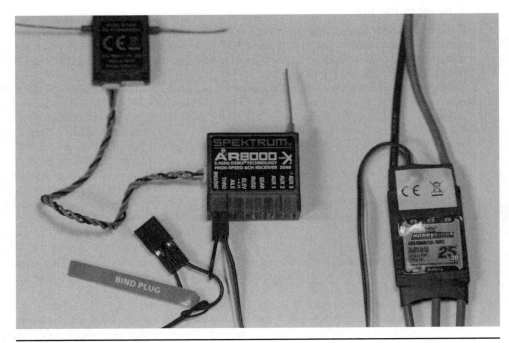

FIGURE 6.8   AR8000 bind plug.

FIGURE 6.9   DX-8 TRAINER/BIND button.

- Plug the power into the receiver. (I just connected the battery in the experimental setup.) The LED on the AR8000 will start to blink.
- Move the DX-8 throttle to its minimum position for the experimental setup.
- Turn on the DX-8 power, while depressing and holding the TRAINER/BIND button. Figure 6.9 shows this button.
- The AR8000 will then bind with the DX-8. It should take only a few seconds. The LED on the AR8000 will stop blinking and remain on.
- Remove the bind plug from the AR8000. Do not lose this plug. You will need it the next time you have to bind your receiver with a DX-8 transmitter.

NOTE: I strongly advise you to rebind your Elev-8 once you have set it to its final flight configuration. That way, you can be assured that all the fail-safe settings are stored in the AR8000 memory.

Next, I will demonstrate how the DX-8 system functions now that I have taken you through a fairly thorough examination of the theory behind a modern DSSS system.

Experimental R/C System Demonstration

I will be using the same experimental system that I introduced in the last chapter to explore the actual functioning of the DX-8 transmitter and the accompanying AR8000 receiver. I removed the propeller for the initial full-range tests because I wanted to perform a test at a 100% throttle setting, which is not safe to do with the propeller attached.

The complete experimental setup is shown in Figure 6.10. I did bind the DX-8 to the AR8000 so that it would function as needed for the tests.

I also added some test-point connections to two of the ESC leads connecting to the motor so that I would have some place both to monitor the motor power waveforms and to connect

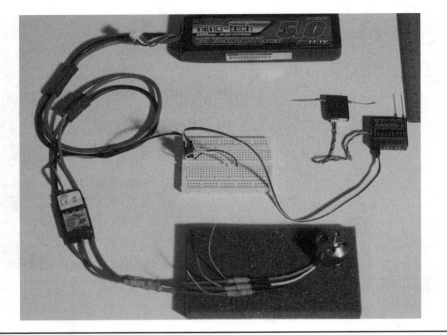

FIGURE 6.10 Actual experimental setup.

the r/min telemetry sensor. Figure 6.11 is a photo of one of these test-point extensions. Notice that I used EC3 connectors throughout, which standardizes all the connection points.

The initial test simply generates the receiver waveform for a zero throttle position. Figure 6.12 shows the waveform out of the AR8000 going to the ESC with the DX-8 throttle at its minimum setting. The automated measurements at the bottom of the waveform show a 1.11-ms high-level pulse width with a 22-ms period, which is equivalent to sending a

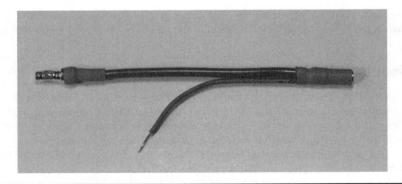

FIGURE 6.11 Test point extension cable.

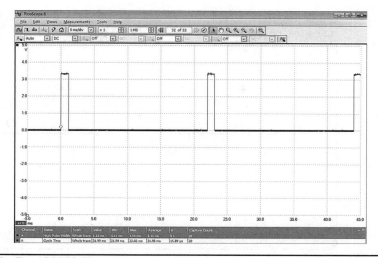

FIGURE 6.12   The AR8000 0% throttle channel output.

0 r/min command to the motor. This is confirmed by my observations that no motor rotations were happening and that there were zero power pulses sent from the ESC to the motor, as shown in Figure 6.13.

The next test generates the waveform for a 100% throttle position. Figure 6.14 shows the waveform out of the AR8000 with the throttle set at 100%. This waveform now shows a 1.88-ms pulse-width again with a 22-ms period, which is the 100% throttle setting. The motor

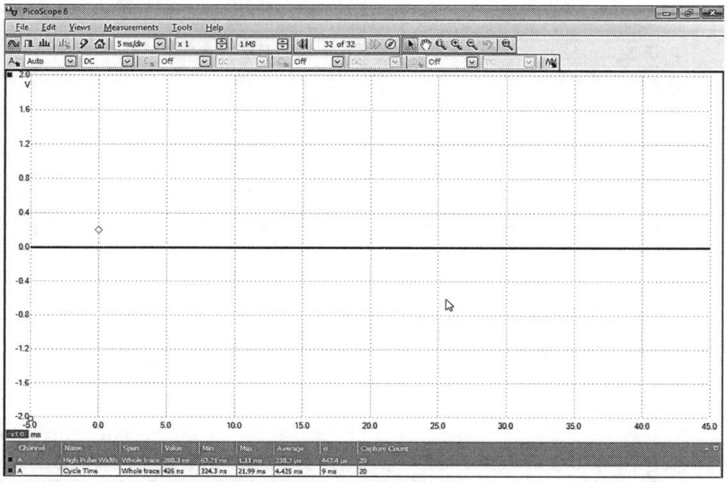

FIGURE 6.13   The ESC-to-motor waveform at 0% power.

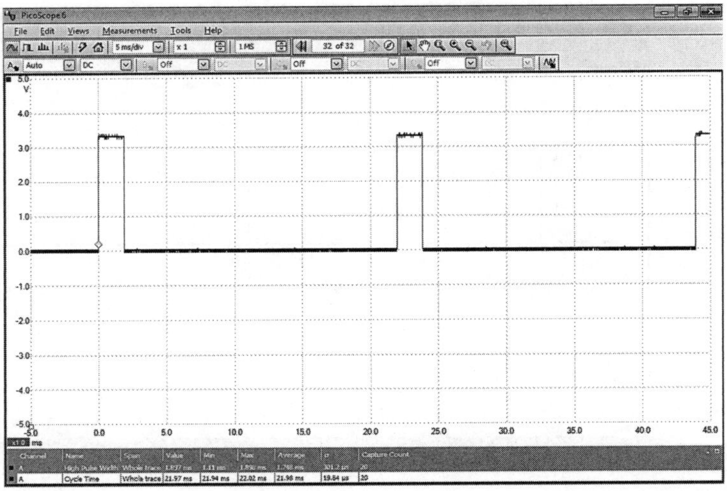

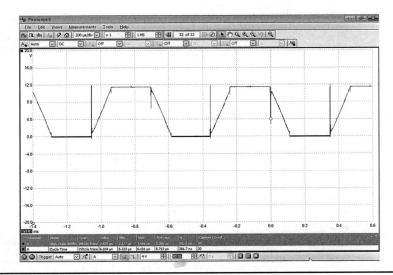

FIGURE 6.14   AR8000 100% throttle channel output.

was observed spinning at a very high rate, and the power pulses from the ESC to the motor were very evident, as shown in Figure 6.15.

This 100% power-pulse waveform is remarkably clear with just a few switching transient pulses obvious in the waveform. You can easily view the up and down ramps that are on either side of the power pulse. I discussed how ramping was configured in the last chapter's ESC section but not why it is needed. Ramping is required to allow sufficient time for the *electromagnetic* (EM) fields located at the stator poles to both build and decay. A synchronized rotating EM field cannot be established and maintained without allowing enough transition time, which is the reason for the ramps.

FIGURE 6.15   ESC motor waveform at 100% power.

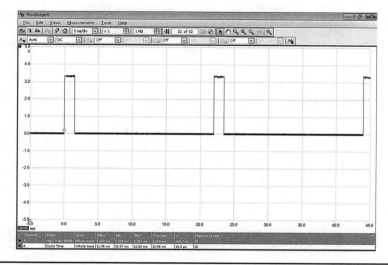

Figure 6.16   The AR8000 50% throttle channel output.

I also captured the waveforms at the throttle 50% setting, which was set using the marked midpoint of the throttle control. Figure 6.16 shows a pulse width of 1.50 ms and 22-ms period.

The corresponding power pulses for the 50% setting are shown in Figure 6.17. Notice that it is considerably noisier than the 100% power waveform. I am not sure why this is the case, but it may be due in part to a longer cycle time of 800 microseconds (μsec) versus 700 μsec for the 100% power setting.

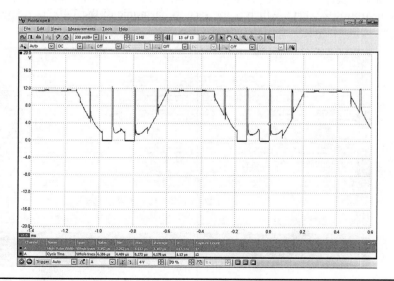

Figure 6.17   ESC motor waveform at 50% power.

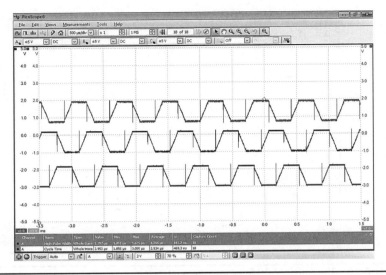

Figure 6.18 Three-motor-lead waveform at 100% power level.

I decided that it would be useful to capture a waveform showing the simultaneous power pulses for all three motor leads. This waveform is shown in Figure 6.18 with the motor operating at a 100% setting.

My careful inspection revealed that all three power pulses repeat every 750 μsec or approximately 1.333 kHz. I was a bit curious to know if I could relate that frequency to the audible noise emanating from the motor. I downloaded a free app to my smartphone that provided a time versus frequency spectral analysis. Figure 6.19 shows the phone screen with the spectral plot recording the motor noise when the motor is operating at 100% power.

Many spectral lines are present in the plot, including one at about 1.4 kHz. The strongest one is around 2.8 kHz, which is the second harmonic of the 1.4-kHz frequency. From my

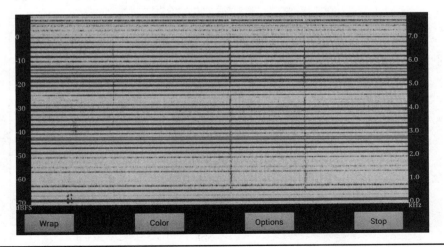

Figure 6.19 Smartphone audio spectral plot.

Channel Name	Position	Pulse width (ms)	Remarks
Aileron	Full left	1.897	22-ms pulse rate
	Full right	1.117	
	Centered	1.499	
Rudder	Full left	1.891	22-ms pulse rate
	Full right	1.105	
	Centered	1.499	
Gear	0	1.580	22-ms pulse rate
	1	1.462	
Aux 1 (labeled as FLAP on DX-8)	0	1.855	22-ms pulse rate
	1	1.462	
	2	1.113	
Aux 2	0	1.899	22-ms pulse rate
	1	1.505	
	2	1.112	
Aux 3	Fully CCW	1.899	22-ms pulse rate
	Center	1.500	DX-8 chirps as the knob is rotated through the center position.
	Fully CW	1.112	

TABLE 6.3   Test Results for Remaining AR8000 Channels

experience as an acoustics engineer, it is not an unusual occurrence to have a second harmonic as the loudest component in an audio spectrum. Of course, the audio noise would become much louder, and the components would change when a propeller is attached to the running motor.

I conducted another pulse-width test on the elevator R/C channel. The DX-8 elevator control stick is on the right side and is spring loaded to stay in a center position. I measured the pulse width to be 1.5 ms when the stick was centered. I then pushed the throttle all the way up, and the width changed to 1.9 ms. Pushing it all the way down produced a 1.1-ms pulse width. The pulses also repeated at a 22-ms rate, which is exactly the same as the throttle control pulses.

The next portion of my experiment was to determine how the rest of the main AR8000 receiver channels functioned with regard to the pulses generated. I tested the remaining channels with the USB oscilloscope and recorded the results as shown in Table 6.3.

Measuring R/C Channel Pulse Width and Rate with the BOE

I realize that most readers will not have the sophisticated USB oscilloscope that I used and may not even have a regular oscilloscope. I highly recommend that you get one if you want to modify your quadcopter or simply to conduct experiments. Many highly capable two-channel units are on the market with a few at, or even below, $400. Readers lacking an oscilloscope can use the following programs written and executed on the BOE to measure both pulse widths and rates. Remarkably, these two programs report measurements that match very closely to those achieved with the USB oscilloscope.

The BOE simply uses a normal servo cable to plug into the selected R/C receiver channel, as shown in Figure 6.20.

FIGURE 6.20   BOE connected to the AR8000 throttle channel.

BOE Pulse-Width Measurements

The first program that I will discuss is named PWM2C_SIGDemo, which measures pulse width. The program has four separate Spin components that are required for it to execute. I am showing you only two of the four, since the other two, while needed, do not function in the actual measurement process. They support text strings and communication with the *Propeller Serial Terminal* (PSerT) program, which is used to display the results.

The essential listing of the PWM2C_SIGDemo program is:

```
CON
  _clkmode = xtal1 + pll16x
  _xinfreq = 5_000_000

  _baudRateSpeed = 250_000
  _newLineCharacter = 13
  _homeCursorCharacter = 1
  _clearToEndOfLineCharacter = 11

  _receiverPin = 31
  _transmitterPin = 30
  _leftServoPin = 14          'changed from 0 in the original for the
                               BOE

  _rightServoPin = 15         '1 in the original
```

```
OBJ
  sig: "PWM2C_SIGEngine.spin"
  com: "RS232_COMEngine.spin"      'not discussed - for serial comm
  str: "ASCII0_STREngine.spin"     'not discussed - for text
                                    manipulation

PUB demo
  ifnot( com.COMEngineStart(_receiverPin, _transmitterPin,
_baudRateSpeed) and { } sig.SIGEngineStart(_leftServoPin,
_rightServoPin, 100))
    reboot

  repeat
    com.transmitString(string("Left: "))
    com.transmitString(1 + str.integerToDecimal(sig.
        leftPulseLength, 10))
    com.transmitString(string(_clearToEndOfLineCharacter,
        _newLineCharacter, "Right: "))
    com.transmitString(1 + str.integerToDecimal(sig.
        rightPulseLength, 10))
    com.transmitString(string(_clearToEndOfLineCharacter,
        _homeCursorCharacter, "Servo Pulse Length's",
        _newLineCharacter))
```

This program is the driver that uses an "engine" object to perform the pulse-width measurement and then reports the results back to the PSerT application for user display. The heart of this program is the forever-repeating loop that sends a series of strings over to the PSerT. The pulse-length values are acquired in the two object-method calls sig.leftPulseLength and sig.rightPulseLength. The sig object is a reference to a program named PWM2C_SIGEngine, which is listed below in an abbreviated manner. I did add some comments to key code lines to help explain what was happening in the code.

```
VAR
  long leftLength, rightLength, stack[7]
  byte cogNumber, leftPinNumber, rightPinNumber, timeoutPeriod

PUB leftPulseLength '' 3 Stack Longs         ' Returns the servo
                                             channel's pulse length in
                                             microseconds.
  return leftLength

PUB rightPulseLength '' 3 Stack Longs         ' Returns the servo
                                             channel's pulse length in
                                             microseconds.
  return rightLength
```

```
PUB SIGEngineStart(leftServoPin, rightServoPin, timeout) '' 9 Stack
                                                            Longs


' Starts up the SIG driver running on a cog. Returns true on
      success and false on failure.
' LeftServoPin - Pin for left channel servo pulse width input.
      Between (0 - 31).
' RightServoPin - Pin for right channel servo pulse width input.
      Between (0 - 31).
' Timeout - The timeout period before zeroing the channel pulse
          lengths in  centiseconds. Between 0 and 100. (Try 10).

  SIGEngineStop
  if(chipver == 1)   'checks Prop chip version number. Should be 1
                                                    for current
                                                    version

    leftPinNumber := ((leftServoPin <# 31) #> 0)      'ensures pin #
                                                    in the range 0 to 31
    rightPinNumber := ((rightServoPin <# 31) #> 0) 'ditto
    timeoutPeriod := ((timeout <# 100) #> 0)      'timeout between 0
                                                    and 100
    cogNumber := cognew(SIGDriver, @stack)      'start a new cog
                                                    with measuring code
    result or= ++cogNumber            'result is a predefined
                                              variable which in this case
                                              stores the cog number


PUB SIGEngineStop '' 3 Stack Longs

' Shuts down the SIG driver running on a cog.

  if(cogNumber)
    cogstop(-1 + cogNumber~)

PRI SIGDriver : leftTimeout | rightTimeout      ' 7 Stack Longs
  ctra := constant(%0_1000 << 26) + leftPinNumber ' sets the cog's
              A counter to start counting as soon as a positive
              edge is detected on the leftPinNumber pin
  ctrb := constant(%0_1000 << 26) + rightPinNumber 'sets the cog's
              B counter to start counting as soon as a positive
              edge is detected on the rightPinNumber pin

  frqa := frqb := 1
  leftTimeout := rightTimeout := cnt

  repeat
```

```
if(phsa < 0)                    ' phsa is a phase lock loop register.
                                It is used to store a count
                                proportional to the high pulse width
  leftLength := 0
  phsa := 0

ifnot(ina[leftPinNumber] or not(phsa))
  leftLength := ((||(phsa~)) / (clkfreq / 1_000_000))
  leftTimeout := cnt

if((cnt - leftTimeout) > ((clkfreq / 100) * timeoutPeriod))
  leftLength := 0

if(phsb < 0)
  rightLength := 0
  phsb := 0

ifnot(ina[rightPinNumber] or not(phsb))

  rightLength := ((||(phsb~)) / (clkfreq / 1_000_000))
  rightTimeout := cnt

if((cnt - rightTimeout) > ((clkfreq / 100) * timeoutPeriod))
  rightLength := 0
```

I would like to point out that no assembly language code was required for this Spin program, since all the timing was accomplished by using the built-in cog counters and

FIGURE 6.21 A Propeller Serial Terminal display of two servo channels.

registers. It is quite possible to resolve timing down to a 12.5-nanosecond (ns) interval by using an 80-MHz system clock with the cog counters. That is impressive measurement accuracy.

Figure 6.21 is a screenshot of the PSerT display with the throttle channel connected to servo pin 14 and the Aux 3 channel connected to servo pin 15.

The DX-8 throttle position was set at 100%, and the Aux 3 knob was at the 50% position for this test. I also connected the USB oscilloscope to the receiver's throttle channel and confirmed that it measured precisely the same value that was displayed on the PSerT screen.

The next program I will discuss concerns the measurement of pulse rates.

BOE Pulse-Rate Measurements

The second program that I will discuss is named jm_freqin_demo, which measures pulse frequency. Pulse frequency is not as important a parameter as pulse width; however, you should measure it to ensure that pulses are arriving at a sufficient rate to constantly update the flight-control board and/or servos. A pulse rate that is too slow could lead to loss of control in the same way that loss of a signal would cause your aircraft to go out of control. An abbreviated listing of jm_freqin_demo is shown below:

```
CON
  _clkmode = xtal1 + pll16x
  _xinfreq = 5_000_000

CON
  #0, CLS, HOME, #8, BKSP, TAB, LF, CLREOL, CLRDN, CR   ' PST format
                                                          control


OBJ
  fc   : "jm_freqin"
  term : "jm_txserial"

PUB main | f

  fc.init(0)                              ' freq cntr on p0

  term.init(30, 115_200)                  ' start terminal
  waitcnt(clkfreq/10 + cnt)
  term.tx(CLS)

  ' setup cog 1 frequency generation
  ' -- note: you may see jitter at high frequencies
  '          due to ctrx pwm behavior

  ' frqx setting = frequency × 2^32 ÷ 80_000_000

  ctra := %00100 << 26                    ' nco/pwm  note: not used
                                            in my version
  ctra[5..0] := 0                         ' use p0
```

```
'frqa := 134_218          ' 2500 Hz
frqa := 3_222             ' 60 Hz
'frqa := 672              ' ~12.5 Hz
dira[0] := 1              ' make p0 output

repeat
  term.str(string(HOME, "Freq: "))
  f := fc.freq                          ' get frequency
  if f > 0                              ' valid?
    term.dec(f/10)                      ' print whole part
    term.tx(".")
    term.dec(f//10)                     ' print fractional part
    term.str(string(" Hz", CLREOL))
  else
    term.str(string("???", CLREOL))
  waitcnt(clkfreq + cnt)
```

This program makes use of another Spin program named jm_freqin, which actually measures and displays the pulse frequency on the Propeller Serial Terminal. The program was created by Jon "JonnyMac" McPhalen (also known as Jon Williams), who is a prolific contributor to the Parallax forums. Jon also made provision for self-generating pulses to test the program. Those pulses emit from pin 0. However, I used pin 14 as an input, since that is one of the servo ports on the BOE. The jm_freqin program is shown below with Jon's introductory comments included because I think they are very helpful in understanding how this program functions:

```
{{
This object uses ctra and ctrb of its own cog to measure the period
of an input waveform. The period is measured in clock ticks; this
value can be divided into the Propeller clock frequency to
determine the frequency of the input waveform. In application, the
period is divided into 10x the clock frequency to increase the
resolution to 0.1Hz; this is especially helpful for low
frequencies. Estimated range is 0.5Hz to ~40MHz (using 80MHz
clkfreq).

The counters are setup such that ctra measures the high phase of
the input and ctrb measures low phase. Measuring each phase
independently allows the input waveform to be asymmetric. In order
to prevent a loss of signal from causing an erroneous value from
the freq() method the fcCycles value is cleared after a valid
frequency is calculated; this means that you should not call this
method at a rate faster than the expected input frequency.
}}

VAR
```

```
          long   cog
          long   fcPin                      ' frequency counter pin
          long   fcCycles                   ' frequency counter cycles

      PUB init(p) : okay
      '' Start frequency counter on pin p
      '' -- valid input pins are 0..27

        if p < 28                           ' protect rx, tx, i2c
          fcPin := p
          fcCycles := 0
          okay := cog := cognew(@frcntr, @fcPin) + 1
        else
          okay := false

      PUB cleanup
      '' Stop frequency counter cog if running

        if cog
          cogstop(cog~ - 1)

      PUB period
      '' Returns period of input waveform

        return fcCycles

      PUB freq | p, f
      '' Converts input period to frequency
      '' -- returns frequency in 0.1Hz units (1Hz = 10 units)
      '' -- should not be called faster than expected minimum input
             frequency

        p := period
        if p
          f := clkfreq * 10 / p             ' calculate frequency
          fcCycles := 0                     ' clear for loss of input
        else
          f := 0

        return f

      DAT

                        org     0

      frcntr            mov     tmp1, par       ' start of  structure
```

```
                    rdlong    tmp2, tmp1              ' get pin#

              mov       ctra, POS_DETECT        ' ctra measures
                                                  high phase
          add       ctra, tmp2
          mov       frqa, #1

          mov       ctrb, NEG_DETECT        ' ctrb measures
                                              low phase
          add       ctrb, tmp2
          mov       frqb, #1

          mov       mask, #1                ' create pin
                                              mask
          shl       mask, tmp2
          andn      dira, mask              ' input in this
                                              cog

          add       tmp1, #4
          mov       cyclepntr, tmp1         ' save address
                                              of hub storage

restart         waitpne mask, mask              ' wait for 0
                                                  phase
          mov       phsa, #0                ' clear high
                                              phase counter

highphase       waitpeq mask, mask              ' wait for pin
                                                  == 1
          mov       phsb, #0                ' clear low
                                              phase counter

lowphase        waitpne mask, mask              ' wait for pin
                                                  == 0
          mov       cycles, phsa            ' capture high
                                              phase cycles

endcycle        waitpeq mask, mask              ' let low phase
                                                  finish
          add       cycles, phsb            ' add low phase
                                              cycles
          wrlong    cycles, cyclepntr       ' update hub

          jmp       #restart

POS_DETECT      long      %01000 << 26
```

```
NEG_DETECT              long     %01100 << 26
tmp1                    res      1
tmp2                    res      1
mask            res     1                  ' mask for frequency input
                                             pin

cyclepntr       res     1                  ' hub address of cycle
                                             count

cycles          res     1                  ' cycles in input period

                fit     492
```

DAT

This program utilizes both the A and B cog counters as noted in Jon's comments. I will not step through all the assembly language lines as I have done in several previous programs. However, I will mention that using either assembly language or C language routines is the only way to handle very high-speed measurements, such as those made by this program. I have not done it myself, but Jon estimates that this program could measure frequencies as high as 40 MHz, which is incredible considering that the Prop clock rate is only 80 MHz.

The other Spin program that is used in this frequency-measuring application is named jm_txserial, which is Jon's adaptation of a fairly standard Spin program named Full_Duplex. It is a common and recommended procedure to use existing open source code and adapt it to your purposes, as Jon has done in this case. I will not discuss jm_txserial specifically other than to say it is a highly efficient communications program that communicates to the PSerT program to display the main program results.

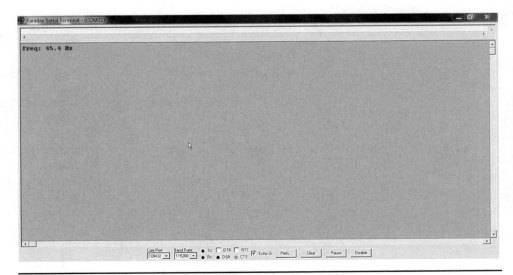

FIGURE 6.22   Measuring the throttle channel pulse frequency.

I connected the AR8000 throttle channel to pin 14 servo port, as was shown in Figure 6.20. I next ran the jm_freqin_demo program and started the PSerT application to view the results. This is shown in Figure 6.22.

The displayed frequency of 45.4 Hz is equivalent to a period of 22 ms, which is computed by taking the reciprocal of the frequency. The 22-ms value exactly matches the pulse rate I measured using the USB oscilloscope. I was impressed with this outcome, since programs often promise certain results but do not deliver on them. That was not the case here; the frequency-measuring capability is very accurate. Just ensure that you keep the maximum input voltage levels at or below 3.3 V, which is the Prop chip's maximum allowable input.

The last section of this chapter deals with telemetry, which you will find very helpful when operating your quadcopter.

Telemetry

The DX-8/AR8000 R/C system has an optional feature that provides telemetry in addition to normal control functions. Telemetry is the automated transfer of data from the aircraft back to the transmitter, which in this case, is also functioning as a receiver. The Spektrum system provides for four data types to be sent via telemetry:

1. Battery voltage
2. Temperature
3. r/min
4. Altitude

Which data type is sent depends on the sensor used to create the initial data. Spektrum utilizes a telemetry module named the TM1000. It is the core of their DSM telemetry system and is shown in Figure 6.23.

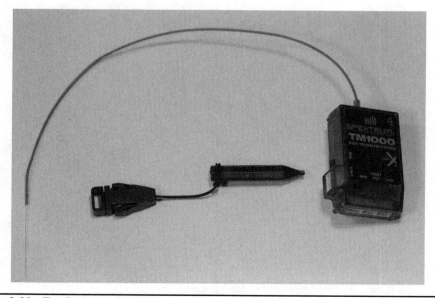

Figure 6.23 The Spektrum TM1000 telemetry module.

Shown alongside the TM1000 module is a plastic bind stick that allows you to press a very tiny button that is visible on the left side of the module. It takes a bit of dexterity during the bind process to power on the receiver-telemetry module and to simultaneously press the bind button during the binding process.

Three ports are visible on the bottom of the module:

1. RPM
2. TEMP/VOLT
3. DATA

The DATA port is committed to link the TM1000 module to the AR8000 receiver where it plugs into the BIND/DAT port. The RPM sensor plugs into the RPM port, and a supplied Y-Connector cable plugs into the TEMP/VOLT port. You will receive a temperature sensor and a voltage sensor as part of the TM1000 package. The RPM sensor is a separately purchased item, and depends on what motor type you are using. Fuel-driven engines use a different sensor from the one used for a BLDC motor. Therefore, I purchased a BLDC sensor. Both the voltage and temperature sensors plug into the Y-connector mentioned above. Figure 6.24 shows the AR8000, TM1000, and the three sensors discussed above.

I connected the sensors to the experimental system and operated the system at about a 50% throttle setting. In Figure 6.25, the DX-8 LCD screen shows the battery voltage, r/min, and temperature.

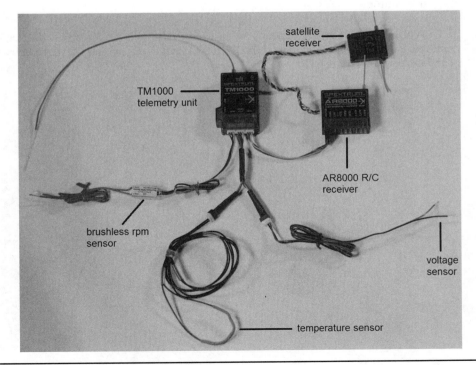

Figure 6.24   The Telemetry system with sensors.

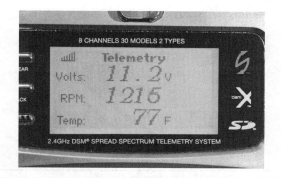

FIGURE 6.25   The DX-8 LCD showing real-time telemetry.

The r/min reading is low because I did not properly configure either the number of poles or the preset ratio in the telemetry setup menu. It should be approximately 4100 r/min instead of the 1216 number shown in the figure. The temperature reading is the ambient temperature around the setup, and the voltage is the LiPo's battery level powering everything.

A four-pin socket labeled as X-BUS is visible in Figure 6.23 next to the BIND switch. Figure 6.26 is a close-up photo of this socket clearly showing the four pins. This socket was designed by Spektrum to allow these additional sensors to be attached to the TM1000:

- G-Force three-axis accelerometer (low range—up to 8 g's)
- G-Force three-axis accelerometer (high range—up to 40 g's)
- High Current sensor
- Airspeed
- PowerBox
- JetCat

Each of the sensors listed above comes with an extension cable that is plugged into the X-BUS socket. Each sensor also has two sockets allowing for a daisy chain to be formed if

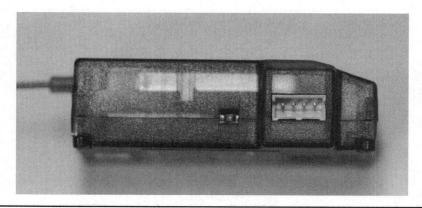

FIGURE 6.26   The X-BUS socket.

FIGURE 6.27 The X-BUS High Current sensor.

more than one sensor is to be used at the same time. Figure 6.27 shows the High Current sensor with the two X-BUS sockets shown on the sensor body. Also, notice that this sensor already comes equipped with EC-3 connectors, which make the connection to the LiPo battery very easy if you use that type of connector.

Spektrum has not published any information regarding the X-BUS; however, it is reported to be an I²C bus as a result of some clever reverse engineering by an R/C aficionado. This bus uses a minimum of three wires, while Spektrum adds another one for some additional functions. The four connections are:

1. SDA—Serial data
2. SCL—Clock line
3. UBatt—Proprietary Spektrum function
4. GND—Ground

The Inter-Integrated Circuit interface or I²C (pronounced eye-two-cee or eye-squared-cee) is also a synchronous serial data link. Figure 6.28 is a block diagram of the I²C interface

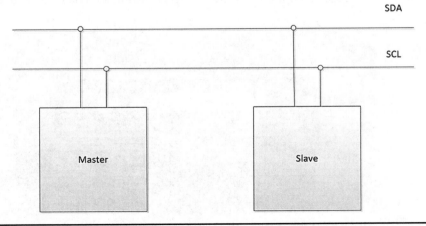

FIGURE 6.28 The I²C block diagram.

FIGURE 6.29   The X-BUS extension cable.

showing one master and one slave. This configuration is also known as a multidrop or bus network.

The I²C supports more than one master as well as multiple slaves. This protocol was created by the Philips Company in 1982 and is a very mature technology, meaning it is extremely reliable. Only two lines are used: SCLK for serial clock and SDA for serial data. You will need to purchase and cannibalize the extension cable shown in Figure 6.29 to get access to the X-BUS.

The only caution I would add, if you are considering using the I²C bus to transfer data, is to remember that it is a low-speed bus. The I²C bus can operate at up to 400 kHz, but it was determined that the X-BUS operates only at 100 kHz. That seems fast, but it is actually fairly slow, when considering the volume of data that some sensors generate.

This last section on telemetry wraps up this chapter. The next chapter examines R/C servos that are used extensively in R/C aircraft. While the basic Elev-8 does not use servos per se, it would be in your best interest to learn about servos and how to best incorporate them into an R/C system.

Summary

I began this chapter with a brief history of R/C development and its rapid progression after the transistor was invented.

Next, I discussed the various modulation schemes used with R/C systems, focusing on PPM and PWM modes. I also discussed the importance of counteracting noise, which is the main culprit in loss-of-control situations. I compared how the various modulation types cope with noise and showed that AM is the worst, FM is somewhat better, and finally, that DSSS is the best.

DSSS was next discussed in depth because it is a superior modulation technique and is the type used in my Elev-8 R/C system. The FHSS was also mentioned, since it is the major competitive R/C modulation scheme used in 2.4-GHz systems.

Then I discussed the binding process between Spektrum's DX-8 transmitter and the AR8000 receiver. Binding must be done before a DSSS system is used because certain transmitter data must be stored in the receiver's EEPROM. I explained the reasons for this in the DSSS write-up.

The next section dealt with an experiment that was designed to demonstrate how the R/C transmitter and receiver function using PWM. I used a slightly modified setup from the experiment shown in Chapter 5. Several USB oscilloscope screenshots showed the pulse waveforms associated with 0, 50, and 100 percent throttle settings. I also discussed all the remaining AR8000 channels showing that they operated in a similar fashion to the throttle channel.

A demonstration of how the Propeller BOE could be used to measure pulse width was shown along with the majority of the program code. The BOE and USB oscilloscope results were practically identical.

Another BOE program was run to show you how to directly measure pulse frequency. You determine the pulse rate by calculating the reciprocal of the frequency. A measure of the pulse frequency or rate is an important number because it ensures that your R/C system is updating at a fast enough rate. An update that is too slow could lead to loss of control.

The chapter concluded with a detailed discussion of the optional Spektrum telemetry system. I showed you all of the standard sensors and discussed some of the more advanced sensors. I also discussed Spektrum's X-BUS, which in reality, is the standard I²C serial bus. I mentioned that this bus could be used to add your own sensors and even microcontrollers onto the telemetry bus.

CHAPTER 7

Servo Motors and Extending the Servo Control System

Introduction

In this chapter, I will show you what makes up a standard R/C servo motor, as well as how to control one. I have already provided a considerable amount of information in the three previous chapters on how pulse signals are used to control servos. Now it is time to reveal the inner workings of a servo motor so that you will understand how and why it operates as it does and be aware of its limitations and constraints when you are using it. This "reveal" will focus on how a specific pulse width translates into a specific servo-motor motion.

I will also discuss how a standard servo motor can be converted into a *continuous rotation* (CR) servo motor. CR motors operate a bit differently than standard ones do. The CR motor has the pulse width that directly controls the continuous angular speed or rotation instead of providing a limited angular motion as the standard servo motor does. CR servo motors are often used as replacements for conventional motors in which low torque requirements exist, such as for powering small R/C cars or boats. I always use CR servos to power my robotic projects, and they seem to function quite nicely.

The next section describes how I built a system to measure the pulse widths for up to three of the R/C channels. I will also show you how to display the results on a 4 × 20 LCD character display. This system uses the Parallax *Board of Education* (BOE) and can be made totally portable by powering it all from a standard 9-V battery. My discussion of the software includes quite a bit of information regarding pointers and indirection, which are often a source of confusion for beginning programmers. Also, in the software, I point out how a Spin program measures pulse width in a way that has not been previously shown in this book.

This chapter concludes with a discussion on two ways to extend the standard servo-control system to accomplish functions that enhance the Elev-8 platform. The first one controls the onboard LED lighting strips mounted on the bottom of the Elev-8 booms. The second one controls a tilting mechanism for a remote-controlled, *first-person viewer* (FPV), which can be attached to the bottom of the Elev-8. The actual FPV will be described in a later chapter. Now, I want to focus on only the servo-control aspect of this system.

Exploring a Standard R/C Analog Servo Motor

Figure 7.1 is a partially transparent view of the inner workings of a standard R/C servo motor. I would like to point out five components in this figure:

Figure 7.1   Inner view of a standard R/C servo motor.

1. Brushed electric motor—left side
2. Gear set—just below the case top
3. Servo horn—attached to a shaft protruding above the case top
4. Feedback potentiometer—at the bottom end of the same shaft with the horn
5. Control PCB—bottom on the case to the motor's right

The electric motor is just an inexpensive, ordinary motor that probably runs at approximately 12,000 r/min unloaded. It typically operates in the 2.5- to 5-V DC range and likely uses less than 200 mA, even when fully loaded. The servo-torque advantage results from the motor spinning the gear set so that the resultant speed is reduced significantly, which in turn, results in a very large torque increase as compared to the motor's ungeared rating. A typical motor used in this servo class might have a 0.1 oz-in torque rating while the servo-output torque could be about 42 oz-in, which is a 420 times increase in torque production. Of course, the speed would be reduced by the same proportional amount, going from 12,000 r/min to about 30 r/min. This slow speed is still fast enough to move the servo horn to meet normal R/C requirements.

The feedback potentiometer attached to the bottom of the output shaft is a key element in positioning the shaft in accordance with the pulses being received by the servo electronic-control board. You may see the feedback potentiometer clearly in Figure 7.2, which shows another image of a disassembled servo.

Figure 7.2   Disassembled servo showing the feedback potentiometer.

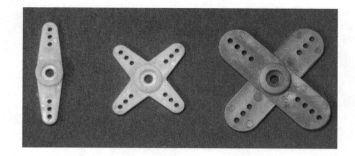

FIGURE 7.3   Servo horns.

I will tell you more about the potentiometer's function during the control-board discussion. For now, I will simply state that it forms part of a closed-loop control system that I introduced in Chapter 2. If you skipped that part, it might be a good time to go back and review the topic, as it applies to this case.

The servo horn is simply a plastic part that slips into grooves at the end of the output shaft and is used as part of a mechanical actuating mechanism. It is held in place with a very small machine screw. The shaft grooves ensure that the horn does not slip under load. Figure 7.3 shows a close-up of some typical servo horns.

The electronics board is the heart of the servo and controls how the servo functions. I will be describing an analog-control version, since that is, by far, the most popular type used in low-cost servo motors. I will mention the digital version at the end of this section and compare it to the analog version. Figure 7.4 shows a Hitec control board that is in place for the model HS-311 that I used for my demonstration system.

The main chip is labeled HT7002, which is a Hitec private model number as well as I could determine. This chip functions in the same way as a commercially available chip by Mitsubishi with the model number M51660L. I will refer to the M51660L in my discussion, since it is used in a number of other manufacturer's servo motors and would be representative of any chip that is used in this situation. The Mitsubishi chip is entitled "Servo Motor Controller for Radio Control," and its pin configuration is shown in Figure 7.5.

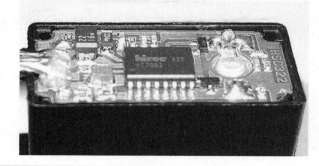

FIGURE 7.4   Hitec HS-311 electronics board.

PIN CONFIGURATION (TOP VIEW)

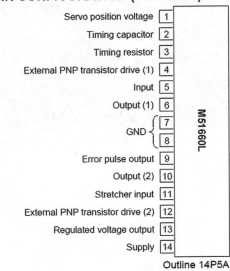

Servo position voltage — 1
Timing capacitor — 2
Timing resistor — 3
External PNP transistor drive (1) — 4
Input — 5
Output (1) — 6
GND { 7 / 8 }
Error pulse output — 9
Output (2) — 10
Stretcher input — 11
External PNP transistor drive (2) — 12
Regulated voltage output — 13
Supply — 14

M51660L

Outline 14P5A

FIGURE 7.5   Mitsubishi M51660L pin configuration.

Don't be put off by the different physical configuration between the HT7002 in Figure 7.4 and the chip outline in Figure 7.5, as it is often the case that identical chip dies are placed into different physical packages for any number of reasons. The M51660L block diagram shown in Figure 7.6 illustrates the key functional circuits incorporated into this chip.

Now I will provide an analysis that will go hand-in-hand with the illustration of the demonstration circuit shown in Figure 7.7, which was provided in the manufacturer's data sheet (as were the previous two figures 7.4 and 7.5).

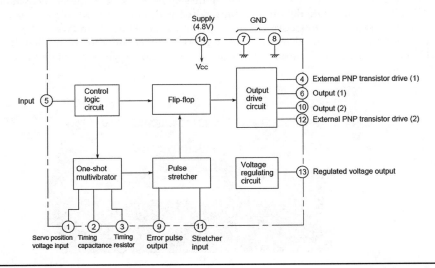

FIGURE 7.6   M51660L block diagram.

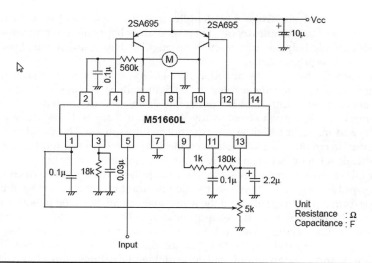

FIGURE 7.7   Demonstration M51660L schematic.

The analysis below should help you understand how an analog servo functions and why there are certain limitations inherent in its design.

1. The start of a positive pulse appearing on the input line (pin 5) turns on the *set/reset* (RS) flip-flop and also starts the one-shot multivibrator running.
2. The RS flip-flop works in conjunction with the one-shot to form a linear one-shot, or monostable, multivibrator circuit whose "on-time" is proportional to the voltage appearing from the tap on the feedback potentiometer and the charging voltage from the timing capacitor attached to pin 2.
3. The control logic starts comparing the input pulse to the pulse being generated by the one-shot.
4. This ongoing comparison results in a new pulse called the error pulse, which is then fed to the pulse-stretcher, deadband, and trigger circuits.
5. The pulse-stretcher output ultimately drives the motor control circuit that works in combination with the directional control inputs that originate from the RS flip-flop. The trigger circuits enable the PNP transistor driver gates for a time period directly proportional to the error pulse.
6. The PNP transistor driver-gate outputs are pins 4 and 12, which control two external PNP power transistors, which can provide over 200 mA to power the motor. The M51660L chip can provide only up to 20 mA without using these external transistors. That is too little of a current flow to power the motor in the servo. The corresponding current *sinks* (return paths) for the external transistors are pins 6 and 10.
7. The 560-kΩ resistor (R_f), connected between pin 2 and the junction of one of the motor leads and pin 6, feeds the motor's back *electromotive force* (EMF) voltage into the one-shot. Back EMF is created within the motor stator winding when the motor is coasting or when no power pulses are being applied to the motor. This additional voltage input results in a servo *damping effect*, meaning that it moderates or lessens any servo overshoot or in-place dithering. I will also further discuss the R_f resistor when I cover the CR servo operation.

The above analysis, while a bit lengthy and detailed, was provided to give you an understanding of the complexity of what is constantly happening within the servo case. This knowledge should help you determine what might be happening if one of your servos starts operating in an erratic manner.

The word *deadband,* as mentioned in step 4 of the analysis, is worth further explanation. Deadband used in this context refers to a slight voltage change in the control input that should not elicit an output. This is a deliberate design feature created for the instance when you do not want the servo to react to any slight input changes. Using a deadband improves servo life and makes it less jittery during normal operations. The deadband is fixed in the demonstration circuit by a 1 kΩ-resistor connected between pins 9 and 11. This resistor forms another feedback loop between the pulse-stretcher input and output.

The last servo parameter I will discuss is the pulse-stretcher gain, which largely controls the error pulse length. This gain in the demonstration circuit is set by the values of the capacitor from pin 11 to ground and the resistor connected between pins 11 and 13. This gain would also be referred to as the proportional gain (K_p) in closed-loop control theory. It is important to have the gain set to what is sometimes jokingly called the "Goldie Locks region," not too high and not too low, but just right. Too much gain makes the servo much too sensitive and possibly could lead to unstable oscillations. Too little gain makes it too insensitive and prone to very poor response. Sometimes, experimenters will tweak the resistor and capacitor values in an effort to squeeze out a bit more performance from a servo; however, I believe the manufacturers have already set the component values for a good compromise between performance and stability.

The Digital Servo

It turns out that there are almost no differences between analog and digital mechanical servo components. The mechanical differences, when present, are often related to using metal gears and ball bearings in digital units, which are more expensive than the analog units. However, the main difference is found in the electronic-control board. The analog control was explained in the previous section: analog-control circuits are used in conjunction with digital-logic and comparator circuits. No numeric calculations or *analog-to-digital conversions* (ADC) are done in an analog servo; hence, there is no need for the microcontroller chip that is present in the digital servo.

Figure 7.8 shows three views of a reasonably priced Dynamixel AS-12 digital servo. The ATmega8L servo with its controller board exposed and mounted at the bottom of the servo

FIGURE 7.8 Dynamixel AS-12 digital servo interior views.

Feature	Analog	Digital
Adjust pulse parameters	Unable to adjust; fixed by circuit component values	Able to dynamically adjust for optimal performance
Update frequency or rate	Fixed at the incoming frequency, normally 50 Hz	Receives at 50 Hz but updates to motor at 300 Hz
Deadband	Fixed by component value	Adjustable to suit dynamic operating conditions
Torque	Low to moderate; slow to build to peak value	Moderate to high; very rapid build-up to peak
Power consumption	Low to moderate	Moderate to high
Cost	Low to moderate	Moderate to high

TABLE 7.1   Feature Comparisons between Analog and Digital Servos

can clearly be seen on the right side of the figure. I have already discussed the ATmega8L chip in Chapter 5 because it is the common controller used in many ESCs. In this application, the chip does ADC as well as real-time numeric calculations to generate the appropriate power-control pulse in much the same fashion that the ESC did in response to its PWM signals.

The digital servo has several significant advantages and one big disadvantage as compared to the analog servo, all of which are listed in Table 7.1.

The digital servo outperforms the analog servo in all areas except power consumption, which can be a factor if your aircraft uses many servos. This is actually not that great a concern given the current availability of high-energy and high-capacity LiPo batteries.

I have also included Figure 7.9, which is an excerpt from the Fubata datasheet that shows quite clearly the relationship of the R/C control pulses as they apply to both analog and digital servos.

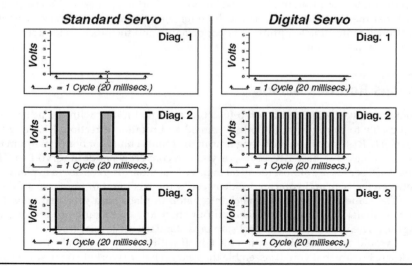

FIGURE 7.9   R/C command pulses applied to both analog and digital servos.

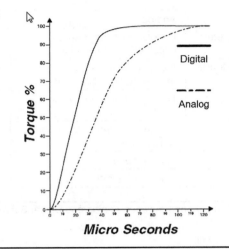

FIGURE 7.10   Deadband characteristic graphs for analog and digital servos.

The analog servo is referred to as the "Standard Servo" in the figure. Also, Diagrams 1, 2, and 3 refer to the no-power, low-power, and high-power operating conditions, respectively. The analog servo is fixed at using the pulses that arrive at the nominal 50-Hz frequency, while the digital servo creates as many as six times that number of analog pulses in the same 20-ms time frame. This means that more average power is being applied to the internal servo motor, which results in more torque and a much faster response. Of course, more average power means more power consumption, which is the digital servo disadvantage.

One more excerpt from the Futaba datasheet is Figure 7.10, which shows a comparison of deadband characteristics, such as percentage of torque versus response time. What you are looking for in these characteristic graphs is a nearly vertical line indicating that the servo rapidly builds torque to the 100% level in a very short time interval. You can see from Figure 7.10 that the digital response is much better than the analog response. This is due to the controller continuing to optimize its performance and the much faster update rate that is in use in the digital unit.

Continuous Rotation Servos

Sometimes you will need a servo to act as a normal motor with the added advantage of being able to closely control both the speed and rotation direction. I have used *continuous rotation* (CR) servos for quite a long time in the robots I build for both classroom and personal use. You may purchase CR servos, or you can convert a standard servo to a CR type fairly easily. CR servos are almost identical in price to standard servos. I will explain the difference between the two, and you can decide if you want to convert or purchase a CR servo.

The standard servo has a mechanical stop in place on a gear that is part of the main output shaft. This tab restricts the output shaft to a fixed range of motion, usually 180°. Figure 7.11 shows this mechanical stop on a standard servo gear train set.

I would recommend snapping the tab off with a sharp diagonal cutter rather than filing it down. Ensure that you disassemble the gear set before working on it, since you don't want

FIGURE 7.11 Mechanical stop.

any plastic shards or filings gumming up the gear train. Figure 7.12 shows the tab neatly removed and filed flat.

The next step in the conversion process is to remove the potentiometer by desoldering it from the circuit board. The potentiometer also has built-in stops, which would restrict the output shaft if it were not removed. The potentiometer must be replaced with a resistor-divider circuit that supplies the midpoint voltage to the one-shot multivibrator. Figure 7.13 shows an altered demonstration schematic with two 2.2-kΩ resistors replacing the potentiometer.

Now the control chip believes it is always at the center point, and when you supply an input-pulse waveform with more than a 1.5 ms width, the controller will drive the motor in a CW direction. Conversely, if the input pulse width is less than 1.5 ms, it will drive the motor in a CCW direction. Additionally, as you either decrease or increase the pulse width, the motor will rotate faster in the respective direction. This means that a 2.0-ms pulse width produces the maximum speed in the CW direction, while a 1-ms pulse width produces the maximum speed in the CCW direction.

The only disadvantage is that the motor will tend to creep if your resistor divider doesn't produce exactly the midpoint voltage. Exactly how much is hard to predict, since the torque

FIGURE 7.12 Tab removed from gear.

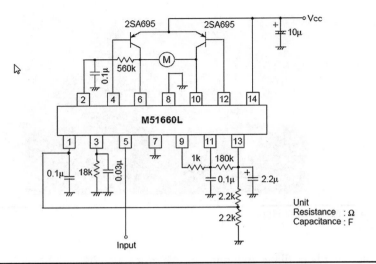

FIGURE 7.13   Demonstration M51660L schematic altered for CR operation.

needed to meet operational requirements plays a part in actually moving whatever object is being powered by the CR servos. A large robot would likely not even move because of the minute creep signal that is created. I would definitely use matched or precision resistors in order to divide the voltage as precisely as possible.

Another way to address this issue is to alter the value of the deadband resistor (the 1 kΩ) to help eliminate the undesired motion. It would take a trial-and-error process to determine the correct value.

There is one final caveat that you should know. It is entirely possible that the plus or minus .5-ms deviation from the center 1.5-ms pulse width will not produce the full rotation speed change that is possible. This is entirely due to having too large a value for the feedback resister R$_f$. The value set for this resistor in the demonstration circuit is 560 kΩ. This may have to be lowered to 120 kΩ to achieve the full speed capability for pulse widths that range from 1 to 2 ms.

NOTE: This conversion process of changing standard servos to CR servos is applicable only to analog servos. I am not aware of any process to convert a digital servo. It may be possible but certainly would involve changing proprietary firmware, which is just not feasible.

R/C Signal Display System

It is critical to confirm the quality and values of pulse signals when designing a control system based upon those signals. The following system uses a BOE and an LCD to display three R/C channel pulse widths in real time. I used an ordinary 9-V battery to power the system in order to show that it has minimal power requirements. Figure 7.14 shows a pictorial diagram of the main components and how they are interconnected using standard three-wire servo cables.

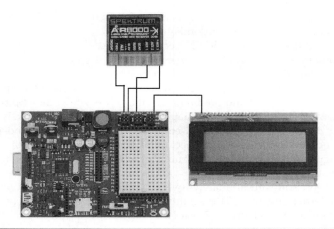

FIGURE 7.14 Pictorial diagram of the real-time, servo-pulse-monitoring test system.

In Figure 7.15, the actual test setup is shown running with three R/C channels being displayed on the LCD. I will discuss the displayed data shortly.

The LCD display is an interesting peripheral. It uses a standard 4 × 20 character display with backlight control to help with the character visibility in various ambient lighting conditions. Normally, LCD displays are parallel devices, which means that they require a total of 8 to 16 control lines from a microprocessor to display data, depending on whether they are in a nibble (4 lines) or byte (8 lines) mode. The LCD I used for this setup has a

FIGURE 7.15 Test system running and displaying real-time data.

Parallax-developed serial-to-parallel "back-pack" auxiliary board, which is shown attached to the back of the main LCD board in Figure 7.16.

This board uses only a single TTL serial line to accept data and display it on the LCD. I used a standard servo-control cable to connect it to the BOE. The secret to this simplified operation is the driver software that is discussed below.

The software running on the BOE is a modified version of a Spin program named RX_Demo. It was created and posted on Parallax's website in their Spin software exchange they call OBEX. This site is a very valuable resource where you will likely find programs that will either directly match your requirements or need only slight modifications to do so. I slightly modified the original top object to take advantage of the built-in servo ports in the BOE configuration. I also reduced the number of R/C channels monitored from six to three, as that satisfied my requirements.

The project software ultimately involved eight Spin files with four of the eight filling what I will term utility roles. These utility files handled the LCD display, serial interface, and numeric conversions. Figure 7.17 is a PSerT screenshot of the beginning of the RX_demo program.

Please notice the Spin program hierarchy shown in the upper left-hand corner of this figure. You can easily see the relationships between the various objects. Essentially, the program named Debug_Lcd takes care of all the display functions needed in the RX_demo program. The RX program does the actual pulse-width detection and reports the results back to RX_demo. Finally, the Servo32v6 program handles any servo pulse modifications that are needed before being sent to the designated output pins.

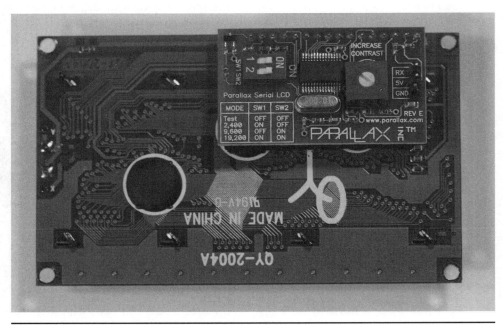

Figure 7.16 LCD display serial-to-parallel converter board.

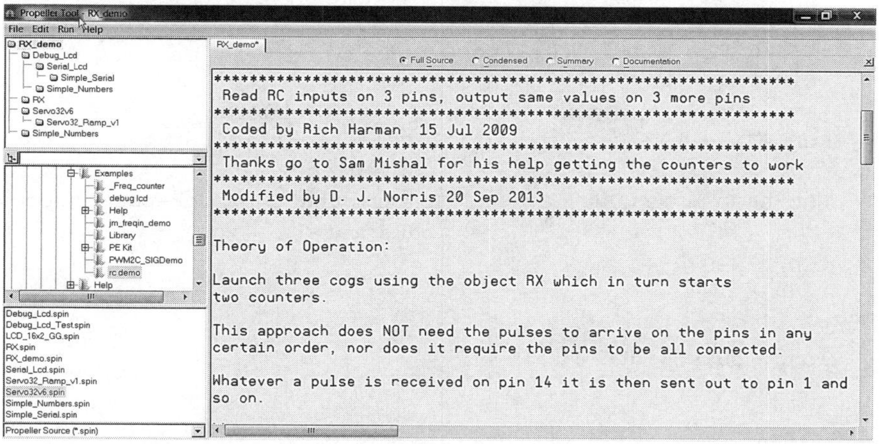

FIGURE 7.17   Start of the RX_demo program in the PSerT.

Below is the RX_demo program listing for which I have included the original documentation as well as my own comments to help clarify the program's functioning.

```
{{
******************************************
* RX Demo version 1.4                                            *
* Author: Rich Harman                                            *
* Copyright (c) 2009 Rich Harman                                 *
* See end of file for terms of use.                              *
******************************************************************
 Read RC inputs on 3 pins, output same values on 3 more pins
******************************************************************
 Coded by Rich Harman  15 Jul 2009
******************************************************************
 Thanks go to Sam Mishal for his help getting the counters to work
******************************************************************
 Modified by D. J. Norris 20 Sep 2013
******************************************************************
Theory of Operation:

Launch three cogs using the object RX which in turn starts two
counters. This approach does NOT need the pulses to arrive on the
pins in any certain order, nor does it require the pins to be all
connected. Whatever pulse is received on pin 14 is then sent out to
pin 1 and so on.
}}
```

```
CON
  _clkmode = xtal1 + pll16x
  _xinfreq = 5_000_000

  LCD_PIN = 19                        'I selected this servo port for
                                       convenience
  LCD_BAUD = 19_200                   'Ensure the LCD back-pack matches this
                                       baud rate
  LCD_LINES = 4

VAR
  long  pulsewidth[4]

DAT
  pins  LONG 14, 15, 16

OBJ
  lcd        : "debug_lcd"
  RX         : "RX"
  servo      : "Servo32v6"
  num        : "Simple_Numbers"



PUB Init
  if lcd.init(LCD_PIN, LCD_BAUD, LCD_LINES)  'init returns True if
                                              it ran

    lcd.cursor(0)                     'cursor set to off
    lcd.backLight(True)               'turn on LCD backlight
                                       for easier character
                                       display

    lcd.cls                           'clear the LCD screen
    lcd.str(string("RX Demo v0.1"))   'welcome banner
    waitcnt(clkfreq + cnt)            'wait for 1 second

  servo.start                         'start the servo
                                       object


  Rxinput

PUB RXinput | i, pulse[4]

  lcd.cls
  RX.start(@pins,@pulseWidth)              'start three RX objects using
                                            the pins array values. RX populates
                                            the pulseWidth array with the data
```

```
        waitcnt(clkfreq/2 + cnt)                    'wait .5 seconds

    repeat
      repeat i from 0 to 2
        pulse[i] := pulsewidth[i]                   'capture pulse values from
                                                    pins 14 to 16
          waitcnt(clkfreq / 2 + cnt)

      updateLCD(pulse[0],pulse[1],pulse[2])  'display pulse values on
                                                    LCD

      out(i, pulse[i])                              'send servo pulses out
                                                    pins 0 to 2

PRI updateLCD(value1, value2, value3) | numstr

    numstr := num.dec(value1)
    lcd.str(numstr)
    lcd.str(string(" "))
    numstr := num.dec(value2)
    lcd.str(numstr)
    lcd.str(string(" "))
    numstr := num.dec(value3)
    lcd.str(numstr)
    lcd.str(string(13))

PUB out(_pin, _pulse)

    servo.set(_pin, _pulse)

DAT
```

When the program runs, the welcome briefly flashes, and then the pulse-width data for three channels is continuously scrolled on the LCD screen. The values are in microseconds, meaning that a value of 1504, as shown on the screen, translates to 1.504 ms.

In the test setup, three of the R/C receiver's channels were connected as follows:

1. Throttle to Servo 14
2. Aux 3 to Servo 15
3. Aux 1 (Flaps) to Servo 16

I then deliberately set each of the corresponding controls on the DX-8 transmitter to its midrange position, which is why you see values near 1500 displayed on the LCD screen in Figure 7.15.

The RX program that measures the incoming pulse width is worth discussing because it uses a different way of determining pulse width than has been previously covered in this book. The code for this is shown below with some clarifying comments after the listing.

```
{{
Theory of Operation:

Launch three cogs using the object RX which in turn each start two
counters. This approach does not need the pulses to arrive on the
pins in any certain order nor does it require all the pins to be
connected. Whatever pulse is received on pin 14 is then sent out to
pin 0 and so on.
}}


VAR
  byte cog[3]
  long stack[60]
  long uS

PUB start(pins_array_address, pulsewidth_array_address)  |  i

  uS := clkfreq/1_000_000

  stop                                     'call the stop method to stop
                                           cogs that may be already started
  repeat i from 0 to 2
     cog[i] := cognew(readPins(@long[pins_array_address][i*2],
  @long[pulsewidth_array_address][i*2]), @stack[i*20]) + 1


PUB stop | i
    repeat i from 0 to 2
      if cog[i]
        cogstop(cog[i]~ -1)

PUB readPins (pins_address, pulsewidth_address) | i, p1, p2,
synCnt, active1, active2

  repeat i from 0 to 1
     spr[8+i]   := %01000 << 26 + long[pins_address][i]    'set the
                                            mode and pin for ctra/b
     spr[10+i] := 1                      'set frqa/b

  p1 := long[pins_address][0]
  p2 := long[pins_address][1]
  dira[p1]~
  dira[p2]~
  long[pulsewidth_address][0]~           'these lines ensure that the
                                         count = 0
```

```
       long[pulsewidth_address][1]~

       active1 := false
       active2 := false

       synCnt := clkfreq/4 + cnt
       repeat until synCnt =< cnt           ' wait 1/4 second to check if
                                              pins are active
         if ina[p1] == 1
           active1 := true
         if ina[p2] == 1
           active2 := true

       repeat
         if active1 == true
           waitPEQ(0, |< p1, 0)             'wait for low state, do not
                                             start counting when high
           phsa~                            'counter set to zero
           waitPEQ( |< p1 , |< p1, 0)       'wait for high
           waitPEQ(0, |< p1, 0)             'wait for low state i.e.
                                             pulse ended
             long[pulsewidth_address][0] := phsa/uS

         if active2 == true
           waitPEQ(0, |< p2, 0)
           phsb~
           waitPEQ( |< p2, |< p2, 0)
           waitPEQ(0, |< p2, 0)
             long[pulsewidth_address][1] := phsb/uS
```

DAT

The RX_demo and RX programs exchange data using a common technique called indirection, where variables represent the physical memory addresses of the data. In C and C++, these indirect variables are called pointers. It is a powerful and very efficient means to exchange data, but be aware that it has inherent dangers, since you can easily misuse a pointer and crash your program and maybe the whole computer. Two references are created in the RX_demo program, as shown below:

```
VAR
  long  pulsewidth[4]
```

```
DAT
  pins  LONG 14, 15, 16
```

The array named pulsewidth has four "long" word elements that may be indexed as pulsewidth[0] to pulsewidth[3]. Data may be written to and/or read from these variable locations. Also, note that it was declared in a VAR block. The memory location variable reference is simply the name pulsewidth prepended with the "@" symbol. So using @pulsewidth tells the program to either store or read data beginning at that location. It is important to state that you should never use an actual physical memory address but simply use the logical reference name or pointer.

The next reference, named pins, also has four "LONG" word elements. Notice that the word "LONG" was capitalized in this declaration. This was a purely optional choice and was done to indicate that the pins array is constant and cannot be overwritten. The pins array was declared in a DAT block, meaning that it is data. It is a read-only data array. You certainly would not want to dynamically change the pin designations for your input channels. However, it can be referred to as pins, just as the pulsewidth array may be referred to as pulsewidth.

The following method call in the RX_demo tells the start method in the RX object where to find the data regarding the input pins and also where to store the pulse-width data corresponding to those pins. Note the indirection used in the arguments.

```
RX.start(@pins,@pulsewidth)
```

The Spin compiler is quite elegant and advanced, since it will automatically create the appropriate reference type based upon the contextual use. The start method signature is shown below:

```
PUB start(pins_array_address, pulsewidth_array_address)
```

The RX.start method call in RX_demo uses @pins and @pulsewidth to pass the starting array addresses. These are copied into the pins_address and the pulsewidth_ address arguments respectively, making them pointers, since they contain addresses not real data. In C and C++, these arguments would have to be separately declared as pointers; however, Spin does it for you as needed. This is a very nice feature.

One more point before I get off my tangent regarding data indirection and pointers. Pointers can be treated as normal data; for example, assigned to other pointer variables or having simple arithmetic operations applied to them. The key point to remember is if you add 1 to a pointer, you are *not* incrementing the physical address but instead are instructing the compiler to use the next logical element in the memory storage area. Using the start method as an example:

$$\begin{aligned} \text{pins\_array\_address} &= \text{start of the pins array} \\ &= \text{address of pins[0]} \\ &= \text{actual value of 14} \end{aligned}$$

Now add 1 to pins_array_address and we get:

$$\begin{aligned} \text{pins\_array\_address} + 1 &= \text{address of pins[1]} \\ &= \text{actual value of 15} \end{aligned}$$

From my teaching experience, I have learned that pointers and indirection have often been difficult concepts for beginners to understand. This is the reason why I presented such a detailed discussion regarding this somewhat complex topic. You really should understand this material in order to get the most from the software development whether you are using Spin, C/C++, or any other language that uses these concepts.

The start method also contains this very complex statement:

```
cog[i] := cognew(readPins(@long[pins_array_address][i*2],
@long[pulsewidth_array_address][i*2]), @stack[i*20]) + 1
```

It is the instruction that creates several cogs to do the pulse-width measurements. This instruction is in a loop that iterates i from 0 to 2. Substituting 0 for i in the first iteration results in this:

```
cog[0] := cognew(readPins(@long[pins_array_address][0],
@long[pulsewidth_array_address][0]), @stack[0]) + 1
```

This really odd expression simply translates as:

> Start executing the method named readPins using the beginning data pointed to by the "pins_array_address" pointer and store the result in the first memory location pointed to by "pulsewidth_array_address." Also, use all the memory needed for the cog located at the beginning of the "stack" memory area.

The index 'i' will increment to 1. Then the whole process will repeat with cog number 1, and it will use the next sequential pin number and store the result in the next sequential location in the pulsewidth array. The stack location is incremented by 20 due to the expression @stack[i*20], which ensures that the new cog has plenty of memory space within which to operate.

The readPins method is the heart of the RX object. I will not step through this method line-by-line other than to point out the liberal use of the waitPEQ instruction in this method. waitPEQ pauses the cog's execution until a pin, which is being monitored, reaches a certain state, normally high or low. Of course, the system counter continues to run, thus accumulating a count directly proportional to the elapsed time. Therefore, the high and low times of a pulse are easily determined using this instruction.

The last portion of the program is concerned with the extended servo outputs that are handled by the Servo32v6 program along with the associated sub-object Servo32_Ramp_v1. This program is a clever extension that will allow you to control up to 32 servos, if you needed such a hefty requirement. I will not be discussing these programs though, since they are not used in either of the two servo applications I discuss below. Just be aware that operating a large number of servos simultaneously can represent a hefty current flow. Even the standard Hitec HS-311 analog servo can take up to 180 mA when operating at no load. The peak-current draw will, of course, go much higher if there is a torque load on the servo. That would mean about an average 3-A current flow, if you were by any chance, trying to run 16 servos at the same time. This is a heavy draw that could rapidly deplete a normal battery.

Elev-8 LED-Lighting Controller

This project is actually a second revision of an LED-lighting controller that I built and installed on my original Elev-8. That controller is shown in Figure 7.18. It worked quite well and was based on the Parallax Basic Stamp II BOE. The LED transistor-driver circuits are located under the hardboard labeled Elev-8 seen on the left side of the figure.

This controller worked correctly as I mentioned above, but I was a bit disappointed that I could not dynamically control the light pattern once it was programmed. I made all the

FIGURE 7.18 Early version of an Elev-8 LED-lighting controller.

LED strips flash in a variety of patterns that just kept repeating, in a way that reminded folks who saw it of the *Close Encounters of the Third Kind* movie without the music.

The more I thought about it, the more I wanted to be able to send a signal to the quadcopter to dynamically change the lighting. One thought was to flash just the LEDs attached to the forward booms so that I could easily see in which direction the quadcopter was moving, especially as daylight faded. I also wanted to stop the flashing entirely in order to conserve battery energy. These ideas would lead to the following requirements:

- Forward boom LEDs flashing
- All LEDs flashing
- No LEDs flashing

These requirements meant that I needed an unused three-state control switch. By serendipity, the switch happened to be available on the DX-8 in the form of the Aux-1, or Flaps, control switch. This light control is an ideal use of the three-position switch because the Elev-8 does not and will never require flaps.

I tested the switch shown in Figure 7.19 and determined that it generated the following pulse widths for the three switch positions, as shown in Table 7.2.

The LED light strips cannot be run directly from the BOE because the current draw is too high. Therefore, I created four transistor driver circuits that control the four strips based on

Switch Position	Pulse Width (ms)
0	1.898
1	1.505
2	1.111

TABLE 7.2 Aux-1 (Flaps) Pulse Width versus Switch Position

FIGURE 7.19   DX-8 Aux-1 (Flap) switch.

gate signals from the BOE, and eventually, from the QuickStart board. As I mentioned earlier in the book, the BOE would be my development platform. I would use it to eventually port all the control programs over to the QuickStart once I have confirmed that everything functions as it should. Figure 7.20 is a picture of the QuickStart board that will be used as the onboard lighting controller.

The following is a Spin program that I wrote to monitor the pulse width being received on the Aux-1 channel, and then, to modify the LED lighting scheme in accordance with the requirements. The transistor-driver circuit is contained in the header documentation; however,

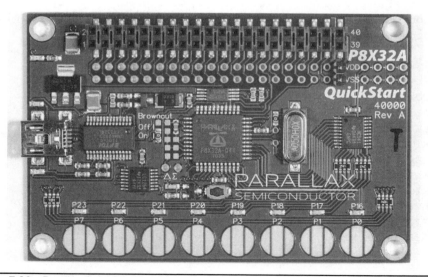

FIGURE 7.20   Parallax QuickStart board.

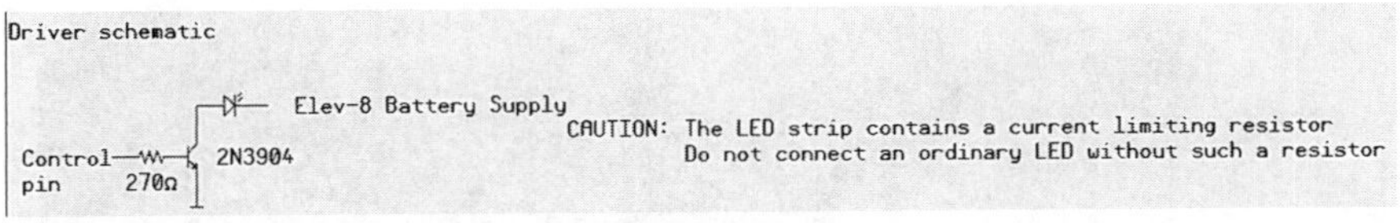

```
Driver schematic


                            ┌─▷┼─  Elev-8 Battery Supply
                            │                CAUTION: The LED strip contains a current limiting resistor
                            │                         Do not connect an ordinary LED without such a resistor
       Control─w──⟨  2N3904
       pin      270Ω │
```

FIGURE 7.21   Transistor driver circuit.

the special characters used to create the schematic do not copy over using the Word program's copy and paste functions. Figure 7.21 shows a screenshot of the schematic portion.

```
{{
LED_Control
D.J. Norris   (C) 2013
This program sets the LED lighting mode for the four LED strips
attached to the underside of the Elev-8 booms. The mode is selected
based upon the Flap switch position on the Spektrum DX-8 R/C
transmitter.
The three possible positions with associated modes and pulse widths
are:

Position              Mode                    Pulse Width
   0          Flash front boom LEDs only        1.899 ms
   1          Flash all LEDs                    1.505 ms
   2          No LEDs are lit or flashed        1.111 ms

Driver schematic

See Figure 7.21, since as already mentioned, Word copy/paste does
not work with the special characters used to create the Spin
documentation schematic.

}}

CON
RIGHT_FRONT = 1
LEFT_FRONT  = 2
RIGHT_REAR  = 3
LEFT_REAR   = 4
WAIT_CNT     = 40_000_000
RIGHT_FRONT_PIN = 5
LEFT_FRONT_PIN  = 6
RIGHT_REAR_PIN  = 7
LEFT_REAR_PIN   = 8

VAR
  long   stack[20]
```

```
PUB init
  dira[RIGHT_FRONT_PIN] := 1
  dira[LEFT_FRONT_PIN]  := 1
  dira[RIGHT_REAR_PIN]  := 1
  dira[LEFT_REAR_PIN]   := 1

PUB start(num)                          'num is the parameter passed from
                                         RX_demo to select the mode

  case num
    0 :
        cognew(mode0,@stack)            'start a fresh cog for
                                         flashing the LED strips

        mode0
    1 :
        cognew(mode1,@stack)
        mode1
    2 :
        cognew(mode2,@stack)
        mode2


    PUB mode0                           'Flashes just the LED strips
                                         on the forward booms

  outa[RIGHT_FRONT_PIN]     := 1
  outa[LEFT_FRONT_PIN]      := 1
  outa[RIGHT_REAR_PIN]      := 0
  outa[LEFT_REAR_PIN]       := 0
  waitcnt(WAIT_CNT + cnt)
  outa[RIGHT_FRONT_PIN]~                'Toggle the pin
  outa[LEFT_FRONT_PIN]~

PUB mode1                               'Flashes all the LED strips
  outa[RIGHT_FRONT_PIN]     := 1
  outa[LEFT_FRONT_PIN]      := 1
  outa[RIGHT_REAR_PIN]      := 1
  outa[LEFT_REAR_PIN]       := 1
  waitcnt(WAIT_CNT + cnt)
  outa[RIGHT_FRONT_PIN]~
  outa[LEFT_FRONT_PIN]~
  outa[RIGHT_REAR_PIN]~
  outa[LEFT_REAR_PIN]~

PUB mode2                               'No LED strips are flashed
  outa[RIGHT_FRONT_PIN]     := 0
  outa[LEFT_FRONT_PIN]      := 0
  outa[RIGHT_REAR_PIN]      := 0
  outa[LEFT_REAR_PIN]       := 0
```

You should realize that the above code is totally useless unless referenced and used by the top object, which is RX_demo. I once again modified RX_demo to use the LED program, and I also added some logic to determine the appropriate LED lighting mode to call. This code listing is abbreviated without the header or license information. I also added comments in an italic font to highlight the LED_Control program changes.

```
CON
    _clkmode = xtal1 + pll16x
    _xinfreq = 5_000_000

    LCD_PIN = 19
    LCD_BAUD = 19_200
    LCD_LINES = 4


VAR
    long  pulsewidth[4]


DAT
    pins  LONG 14, 15, 16, 17


OBJ
    lcd         : "debug_lcd"
    RX          : "RX"
    servo       : "Servo32v6"
    num         : "Simple_Numbers"
    led         : "LED_Control"          'new reference "led"  for
                                         the LED_Control program


PUB Init

if lcd.init(LCD_PIN, LCD_BAUD, LCD_LINES)
    lcd.cursor(0)
    lcd.backLight(True)
    lcd.cls
    lcd.str(string("RX Demo v0.1"))
    led.init
    waitcnt(clkfreq + cnt)

servo.start

Rxinput

PUB RXinput  | i, pulse[4]

  lcd.cls

  RX.start(@pins,@pulseWidth)
```

```
        waitcnt(clkfreq/2 + cnt)

    repeat
      repeat i from 0 to 2
        pulse[i] := pulsewidth[i]              'capture pulse values
                                                from pins 14 to 16

        waitcnt(clkfreq / 2 + cnt)

      updateLCD(pulse[0],pulse[1],pulse[2])    'display pulse values
                                                on LCD

      out(i, pulse[i])                         'send servo pulses out
                                                pins 0 to 2

      if pulse[2] > 1600              'this is the start of the logic to
                                      determine the LED lighting mode
        led.start(0)                  'send a 0 to tell LED_Control to
                                      enter mode0
      if pulse[2] > 1200 AND pulse[2] < 1600
        led.start(1)                           'send a 1 to tell LED_Control to
                                      enter mode1
      if pulse[2] < 1200
        led.start(2)                           'send a 2 to tell LED_Control to
                                      enter mode2

PRI updateLCD(value1, value2, value3) | numstr

  numstr := num.dec(value1)
  lcd.str(numstr)

  lcd.str(string(" "))

  numstr := num.dec(value2)
  lcd.str(numstr)

  lcd.str(string(" "))

  numstr := num.dec(value3)
  lcd.str(numstr)

  lcd.str(string(13))

PUB out(_pin, _pulse)

    servo.set(_pin, _pulse)

DAT
```

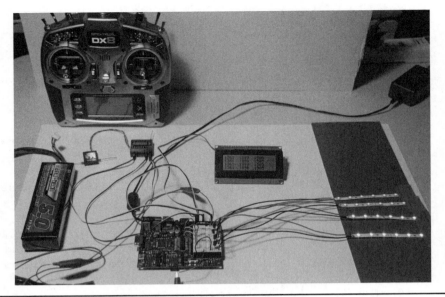

Figure 7.22   The LED-development setup while running in mode 1.

All that you need to do to run this software is to load the new LED_Control and the modified RX_demo into the project, recompile it, and execute it (F11 key). Figure 7.22 shows the LED-development-test setup for the LED-strip control project. I captured all the LEDs being lit as the BOE was operating in mode 1. You may also be able to see that the DX-8 Aux-1 (FLAP) switch is in the middle, or 1, position, which commands that all LEDs flash. Also, note that the LCD display is showing the number 1505 in the rightmost column, which is the Aux-1 pulse width in microseconds.

The actual transistor-switching circuit board that controls the four LED strips and is mounted in the Elev-8 is shown in Figure 7.23.

Figure 7.23   The LED-strip transistor-switching circuit board.

FIGURE 7.24   Installation of the LED-strip transistor-switching circuit board.

The complete transistor-switching board may be installed between the Elev-8 boom ends, as shown in Figure 7.24. I have also wired all the LED-strip power leads to the transistor-switching circuit board in this figure.

Figure 7.25 shows the complete installation of the transistor-switching circuit board along with the QuickStart board. You can see that I used a solderless breadboard to connect all the leads from the transistor-switching circuit board to the QuickStart board as well as to the Aux-1 R/C channel.

I also temporarily connected the LCD screen to verify that the proper pulse signal was being received on the Aux-1 channel. You can see that the pulse width changed very slightly from 1504 to 1505 during the test I ran when the photo was taken. This slight change of one microsecond will not affect the mode selection because the program code uses a much larger test value before a mode change is made. I also deliberately made the decision to use a solderless breadboard to enable quick configuration changes, while recognizing that it is not as reliable as using solid mechanical connections. A critical flight-control problem would not arise if one of these connections were to fail during a flight, since it controls only the LED lighting.

FIGURE 7.25   LED-strip transistor-switching circuit board and QuickStart board installation.

One last thing I want to point out is some of the telemetry data being displayed by the DX-8 LCD screen, as shown in Figure 7.25. The screen shows the LiPo-battery voltage at 12.0 V, the left rear motor rotating at 1293 r/min, and the ambient temperature at 71°. Note that the r/min reading is incorrect for the reasons I previously stated in Chapter 6's telemetry section. It really should be closer to 4300 r/min.

Tilting Mechanism for a First-Person Viewer

The first thing you need to know about this project is that no software is required because it uses only the standard R/C servo-control functionality. I have included this project to show how very simple it is to implement a standard servo actuator to support an enhancement to the Elev-8. The servo will tilt a video camera that is part of a *first-person video* (FPV) system, which is really just a fancy way of saying that there is a video camera mounted on the quadcopter to show you where you are going. I am using the GoPro Hero 3 camera that incorporates both video recording capability and real-time video by using a WiFi connection. Figure 7.26 shows a picture of the Hero 3 camera. I will not say much about it here, since I devote all of Chapter 8 to using video with a quadcopter.

This camera will be attached to the bottom of the Elev-8 with the lens positioned to look forward. This is fine for conducting ordinary flight and avoiding obstacles, such as trees or tall buildings. However, I wanted to increase the camera's flexibility so it would be able to tilt downward and see the terrain and objects beneath the quadcopter while it was either hovering or traveling in a level plane. In addition, I was not concerned about panning the camera, since it is very easy to simply yaw the quadcopter if you want to shift the viewing direction.

After some thought and a bit of research, I came up with a simple tilting-platform design that I am sure has been done before by many others in a similar fashion. Figure 7.27 is a concept sketch that I used before proceeding with the design.

The fixed outer frame is designed for attachment to the Elev-8's bottom chassis plate with some nylon spacers as well as appropriately sized machine screws and nuts. I used Lexan to build the frame and rotatable platform, since it is both strong and easy to work with. I used wood blocks, a bench vise, and a heat gun to bend the Lexan strip to the form shown in the Tilting Platform Assembly drawing, which can be found on this book's website, www.mhprofessional.com/quadcopter.

Figure 7.26   GoPro Hero 3 video camera.

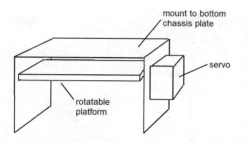

FIGURE 7.27   Concept sketch for the tilting platform.

A Hitec standard HS-311 was used as the table actuator because it seemed to have sufficient torque to turn and hold the camera to the desired position. Figure 7.28 shows the tilt platform without the camera attached to illustrate that it is a simple design.

My only concern with this project was that mounting the camera too far off-center would upset the center-of-gravity, since the assembly with the camera weighs 281 grams. This amount of mass mounted off center could make the quadcopter too unstable to fly. The fully assembled platform assembly with the camera in its water resistant case is shown in Figure 7.29. I mounted the assembly on wooden blocks to provide clearance and allow free camera movement.

The servo cable was attached to the DX-8 Aux-3 R/C channel to test the assembly. I used this control because it creates a continuously variable pulse width from 1.0 to 2.0 ms, which

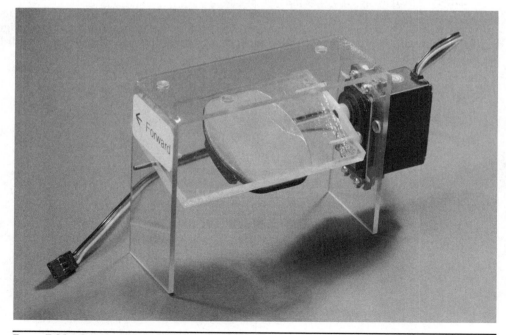

FIGURE 7.28   Camera tilting platform assembly.

FIGURE 7.29   Fully assembled camera platform on a test platform.

means a 0° to 90° range of motion for the rotatable platform. Figures 7.28 and 7.29 show the camera set at 0°, while Figure 7.30 shows the camera set at 90°. This range of motion corresponds to the Aux-3 knob set from fully CCW to fully CW.

If you look carefully at the Figure 7.30, you can see the portion of the BOE in the upper right hand corner that I used to supply power to the servo. The BOE was powered by a 9-V battery, which made for a very portable setup.

The frame for the tiltable platform was attached to the Elev-8 bottom chassis plate, using four 1-in 4-40 screws along with four 5/16-in OD nylon spacers, washers, and nuts. I aligned the frame to the plate and then used a Sharpie™ marker to locate four holes to be drilled

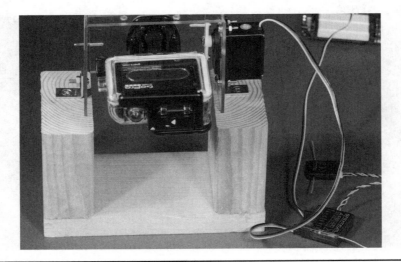

FIGURE 7.30   Camera at the 90° position.

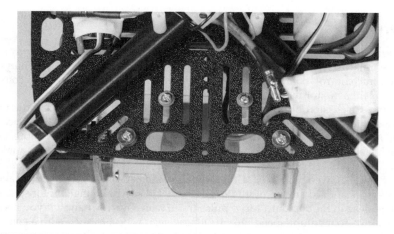

Figure 7.31   Tiltable-frame screw attachments.

through the frame top. Figure 7.31 is a top view of the bottom chassis plate showing the four attachment screw tops. You can readily see the four nylon spacers that support the tiltable frame in Figure 7.32.

My only caution is to ensure that the movable platform that holds the camera is not blocked in its range of motion by any of the mounting screws. I had to slightly notch the platform to allow some clearance for the screws so that the platform could rotate throughout a full 90° range of motion.

This tiltable camera platform project came out very well, and it is much less expensive than any commercial product that will provide a similar function. The total cost of the parts, without the camera, is less than $15.

Figure 7.32   View of the tiltable frame attached to the bottom chassis plate.

Summary

I began this chapter by discussing what is inside a standard analog servo motor and how those innards function. This was followed by a comprehensive circuit analysis of the servo electronic-control board that receives an incoming pulse train and converts it to the equivalent actuator motion.

In a discussion of the digital servo, I pointed out that there was little to no difference between the analog and digital mechanical components. The main difference lies within the electronic-control boards. The digital version provides significantly more torque, and it is much faster at matching the changes in the incoming pulse train than its analog counterpart is.

I next showed you how a *continuous rotation* (CR) servo functions and also how to convert a standard analog unit into a CR unit. CR servos change motor speed and direction in response to the standard servo pulse train, which is very handy if you need a low torque and a low- to medium-speed motor. Otherwise, it is best to stick with a conventional motor that has a complimentary speed control unit attached to it.

Next I discussed a portable servo-signal analysis system that could display on a 4 × 20 LCD screen the pulse widths for up to three R/C channels. The software, which was run on a BOE, was thoroughly analyzed. I also included an in-depth discussion on the subject of indirection and pointers, a sometimes bewildering topic, especially for beginner programmers.

Then I covered two projects, the first of which was an LED-strip lighting controller. This controller is designed to be placed onboard the Elev-8 and controls each LED strip based upon the pulse width sent by the DX-8 Aux-1 (FLAP) channel. There are three separate lighting modes, since the Aux-1 has three positions. This lighting controller enhances the Elev-8 but does not affect its flight performance.

The second project was a tiltable platform that has a GoPro video camera attached to it. The platform is mounted on the Elev-8's bottom chassis plate and tilts to enable the camera to view the ground while it is either hovering or in level flight. The platform is tilted by a standard analog servo that is directly controlled by the DX-8 Aux-3 channel. The camera platform may be continuously tilted from 0 to 90 degrees, since the Aux-3 control is a potentiometer.

The next chapter will show you how to set up and operate an HD real-time video system that uses the tiltable platform to increase the video coverage.

CHAPTER 8

GPS and a Real-Time Situational Display

Introduction

In this chapter, I will discuss a GPS-based location system that is easily carried aloft by an Elev-8 or other quadcopters with similar lifting capacity. The system will transmit its data to a ground station where the quadcopter's position, speed, course, and altitude will be visible on a small display. The coordinates could also be entered into a laptop in order to display the quadcopter's position in the Google Earth application. This system, together with the First-Person Video system described in the next chapter will be used to accurately determine the quadcopter's location and to view the environment in real time.

GPS Basics

We will begin with a short history of the *Global Positioning System* (GPS), and follow that with a detailed explanation of how GPS systems generally function. Then I will focus on the quadcopter's GPS receiver and the development of a real-time display.

Brief GPS History

The GPS is a satellite system that was initially deployed in the early 1970s by the U.S. Department of Defense (DoD) to provide military users with precise location and time synchronization services. Civilian users could also access the system, but its services to civilian users were purposefully degraded by the DoD to avoid any risk that it could be of help to the country's enemies. This purposeful degradation was lifted by order of President Regan in the 1980s to allow civilians full and accurate GPS services.

The current GPS system has 32 satellites in high orbits over the earth. Figure 8.1 shows a representative diagram of the satellite constellation. The satellite orbits have been carefully designed to allow for a minimum of six satellites to be in the instantaneous field of view of a GPS user who is located anywhere on the surface of the earth. A minimum of four satellites must be viewed in order to obtain a location fix, as you will learn in the GPS basics section.

Several other GPS systems are also deployed:

GLONASS—The Russian GPS

Galileo—The European GPS

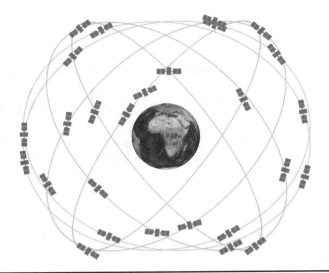

Figure 8.1   GPS system satellite constellation.

Compass—The Chinese GPS

IRNSS—The Indian Regional Navigation Satellite System

I will be using the U.S. GPS system because vendors have made many inexpensive receivers for that system available for purchase. All receivers function essentially in the same way and conform to the *National Marine Electronics Association* (NMEA) standard discussed in a later section.

How GPS Functions

I made up an analogous, fictional, position-location system to help explain how the GPS system functions. First, imagine a two-mile by two-mile land area where this system is set up. The land terrain contains gently rolling hills, each no more than 30 feet in height. The subject, using a special GPS receiver, may be located anywhere within this area. Also located in this area are six 100-ft towers, each containing a beacon. The beacon atop each tower briefly flashes a light and simultaneously emits a loud burst of sound. Each beacon also emits the light and sound pulses once a minute but at a specific time within the minute. Beacon one (B1) emits at the start of the minute, beacon two (B2) at 10 seconds past the start of the minute, beacon three (B3) at 10 seconds later, and so on for the remaining beacons.

It is also critical that the GPS receiver have a line of sight to each beacon and also that the position of each beacon is recorded in an embedded database that is also constantly available to the receiver. The beacon positions B1 through B3 are recorded in x and y coordinates in terms of miles from the origin, which is located in the upper left hand corner of the test area, as shown in Figure 8.2.

The actual position determination happens in the following fashion:

- At the start of the minute, B1 flashes, and the receiver starts a timer that stops when the sound pulse is received. Since the light flash is essentially instantaneous, the time interval is proportional to the distance from the beacon. Since sound

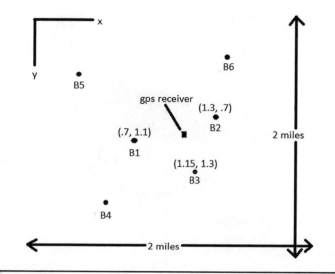

FIGURE 8.2   Beacon test area.

travels nominally at 1100 feet/s (second) in air, a 5-second delay would represent a 5500-ft distance. The receiver must then be located somewhere on a 5500-ft radius sphere that is centered on B1. Figure 8.3 illustrates this abstraction in a graphical representation taken from a Mathworks Matlab application.

- B2 flashes next. Suppose it takes 4 seconds for the B2 sound pulse to reach the GPS receiver. This delay represents a 4400-ft sphere centered on B2. The B1 sphere and B2 sphere are shown intersecting in Figure 8.4. The heavily dashed line represents the

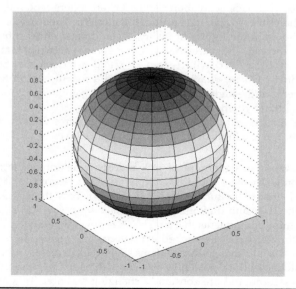

FIGURE 8.3   One sphere.

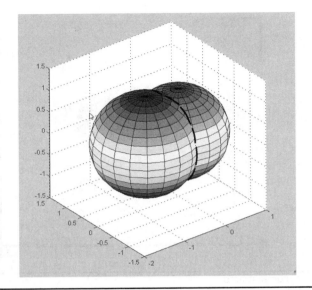

FIGURE 8.4   Two spheres intersecting.

portion of the circle that is the intersection of these two spheres. The receiver must lie somewhere on this circle. The circle is a straight line when observed in a planar or perpendicular view. There is still, however, some doubt or uncertainty as to where the receiver is located on the circle. Thus, another beacon is still needed to resolve the uncertainty.

- B3 flashes next, and suppose it takes 3 seconds for the B3 sound pulse to reach the GPS receiver. This delay represents a 3300-ft sphere centered on B3. The B1, B2, and B3 spheres are shown intersecting in Figure 8.5. The receiver must be located at the star shown in the figure. In reality, it could be at either a high or low point since the third sphere intersects the two other spheres at two points. The receiver position has now been fixed with regard to x and y coordinates but not the third, or z coordinate. Guess what? Now you need a fourth beacon to resolve whether the receiver is at the high or low point. I am not going to go through the whole process again; I think you have figured it out by now.
- Figure 8.5 shows a plane view of all three spheres with the GPS receiver position shown. You can think of it as a horizontal slice taken at $z = 0$, as shown in Figure 8.6.

In summary, it takes a minimum of three beacons to determine the x and y coordinates and a fourth beacon to fix the z coordinate. Now translate beacons to satellites and x, y, and z coordinates to latitude, longitude, and altitude, and you have the basics of the real GPS system.

The satellites transmit digital microwave *radio frequency* (RF) signals that contain both identity and timing components that a real GPS receiver will use to calculate its position and altitude. The counterpart to the embedded database mentioned in my example is called an ephemeris, or celestial almanac, that contains all the data necessary for the receiver to calculate a particular satellite's orbital position. All GPS satellites are in high earth orbits (as mentioned in the history section) and are constantly changing position. This movement requires the receiver to use a dynamic means for acquiring their position fix, which, in turn, is provided

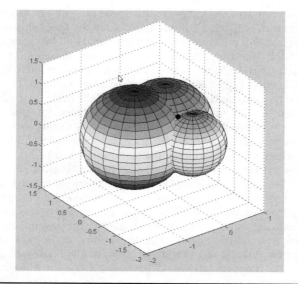

FIGURE 8.5   Three spheres intersecting.

by the ephemeris. This is one reason why it may take a while for a real GPS receiver to establish a lock, as it must go through a considerable amount of data calculations to determine actual satellite positions within its field of view.

In my example, the radii of the "location spheres" are determined by the receiver using extremely precise timing signals contained in the satellite transmissions. Each satellite contains an atomic clock to generate these clock signals. All satellite clocks are constantly synchronized and updated from earth-based ground stations. These constant updates are needed to maintain GPS accuracy, which would naturally degrade because of two relativistic effects. The best way to describe the first effect is to retell the paradox of the space-travelling twin.

Imagine a set of two twins, (male, female—doesn't matter) one of whom is slated to take a trip on a fast starship to our closest neighboring star, Alpha Centauri. This round trip will

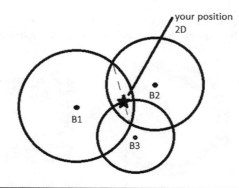

FIGURE 8.6   Plane view of the three spheres intersecting.

take about ten years travelling at nearly the speed of light. The other twin will stay on Earth awaiting the return of his/her sibling. The twin in the space ship will accelerate very close to light speed and will patiently wait the ten years it will take, according to the clock in the ship, to make the round trip. According to Professor Einstein, if the travelling twin could view a clock on Earth he/she would observe time going by more quickly than it did in the spaceship. This effect is part of the *theory of special relativity* and, more specifically, is called *time dilation*. If the twin on Earth could view the clock in the spaceship, he/she would see it turning at a much slower rate than the earthbound clock. Imagine what happens when the travelling twin returns and finds that he/she is only ten years older but that the earthbound twin is 50 years older due to time dilation. The space twin will have time travelled a net 40 years into Earth's future by taking the ten-year space trip!

The second effect is more complex than time dilation. I will simply state what it is. According to Einstein's *theory of general relativity*, objects located close to massive objects, such as the earth, will have their clocks moving slower as compared to objects that are further away from the massive objects. This effect is due to the curvature of the *space-time continuum* predicted and experimentally verified by the general relativity theory.

Now back to the GPS satellites that are orbiting at 14,000 *kilometers per hour* (km/h), while the earth is rotating at a placid 1,666 km/h. The relativistic time dilation due to the speed differences is approximately −7 μsec/day, while the difference due to space-time is +45 μsec/day for a net satellite clock gain of 38 μsec/day. While this error is nearly infinitesimal on a short-term basis, it would be very noticeable over a day. The total daily, accumulated error would amount to a position error of 10 km or 6.2 miles (mi), essentially making GPS useless. That is why the earth ground stations constantly update and synchronize the GPS satellite atomic clocks.

NOTE: *As a point of interest, the atomic clocks within the GPS satellites are deliberately slowed prior to being launched to counteract the relativistic effects described earlier. Ground updates are still needed to ensure that the clocks are synchronized to the desired one nanosecond of accuracy.*

Quadcopter GPS Receiver

I selected the Parallax PMB-688 GPS receiver, which is small, lightweight, and very suitable for use in this project. Figure 8.7 is a picture of this receiver. The PMB-688 GPS receiver will track up to 20 satellite channels, which provides for both fast acquisition and the continuous lock of NMEA data from the satellites. Table 8.1 lists key specifications and features for this receiver.

Several key specifications are worth discussing a bit more. An acquisition sensitivity of −148 dBm means that the receiver is extremely sensitive to picking up weak GPS signals.

FIGURE 8.7   Parallax PMB-688 GPS receiver.

Features/Specifications	Description
Sensitivity	Acquisition: −148 dBm Tracking: −159 dBm (These are very sensitive levels.)
Chipset	SiRFstar III
Channels	20, simultaneous tracking
Data protocol	NMEA 0183 v2.2 GGA, GSV, GSA, RMC (optional VTG, GLL)
Power	Typical 65 mA @12 V (Chip uses 3.3 V to 5 V)
Antenna	Internal patch with provision for external connection
Storage	Rechargeable battery stores real-time clock (RTC) data and receiver configuration settings
Connections	Premade cable with connector for power and data interconnections
LED functions	Power on/off and navigation
Start time	30 seconds

TABLE 8.1   PMB-688 Features and Specifications

The −159 dBm tracking sensitivity means that the signal, once acquired, can lose up to 90% of its original strength, yet remain locked in by the receiver.

Having an NMEA 0183 output operating at 9600 baud means that the receiver generates standard GPS messages at a rate twice as fast as comparable receivers. The 30-second start-up time is excellent and due in part to the receiver's extreme sensitivity.

GPS Receiver UART Communications

Universal Asynchronous Receiver Transmitter (UART) is the serial data protocol used between the GPS receiver and the Propeller Mini processor module (which is discussed in a later section). Three data pins are the minimal amount necessary to establish a communications link between the receiver and the processor. They are identified on the GPS as TTLTX (transmit), TTLRX (receive), and GND (ground or common), as shown in Figure 8.8.

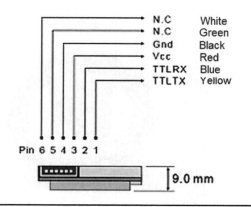

FIGURE 8.8   UART pins.

The TTL in the pin designations represents the fact that the logic levels are 0 and 5 V for low and high levels respectively. The GPS receiver also uses a 9600-baud rate to communicate with the controlling microprocessor to both receive and transmit data back and forth. There is no need for a separate clock signal line, since the UART protocol is designed to be self-clocking.

CAUTION: *To ensure communications with the Prop Mini module, connect the GPS TX lead to the Mini's P8 pin, and likewise, connect the GPS RX lead to the Mini's P9 pin. Misconnecting these pins will likely not cause any damage, but you will not have data communications between the GPS receiver and the Propeller Mini module.*

Initial GPS Receiver Test

It would be wise to check that the PMB-688 GPS receiver is functioning as expected before going on to later stages in this project. Ensure that you have a good line of sight with the open sky to be able to receive the GPS satellite signals. I used an external GPS antenna because my test setup was indoors and had no reliable satellite reception. Parallax has an external GPS antenna available (part number 28502) that is shown in Figure 8.9 and is well worth the modest cost. Erratic or unreliable satellite reception will quickly cause this project to fail.

An interconnecting cable between the GPS and the monitoring laptop will also be needed along with a very useful Prolific USB-to-Serial software driver. The link is set up using a USB-to-TTL serial cable that is connected to the GPS receiver cable, as shown in Figure 8.10. This cable is available from Adafruit Industries as part number 954.

The USB/TTL cable has four pin connectors that are color coded and attached to the matching GPS receiver's color-coded pin connectors, as detailed in Table 8.2. The physical solderless breadboard connections between the GPS receiver cable and the USB/TTL cable are shown in Figure 8.11.

I used the *Propeller Serial Terminal* (PSerT) program with the baud rate set to 4800 to match the GPS receiver output. Additionally, COM port 44 was automatically assigned on the laptop by the Prolific software driver. Figure 8.12 is a screen capture of the GPS data stream showing that the GPS receiver was properly functioning and receiving good satellite signals.

Completing the above steps confirms that the PMB-688 GPS receiver is operating properly, which is a prerequisite before further project development. You are almost ready to

FIGURE 8.9 External GPS Antenna (PMB-688).

FIGURE 8.10   USB-to-TTL serial link cable.

GPS Receiver Cable Color	USB/TTL Cable Color	Function
Black	Black	Ground
Blue	White	TXD (out of laptop)
Yellow	Green	RXD (into laptop)
Red	Red	5 V DC

TABLE 8.2   GPS Receiver to USB/TTL Cable Connections

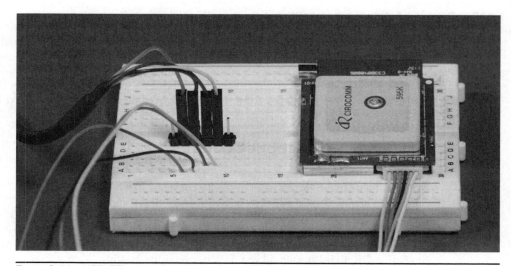

FIGURE 8.11   USB/TTL cable connection from GPS to laptop.

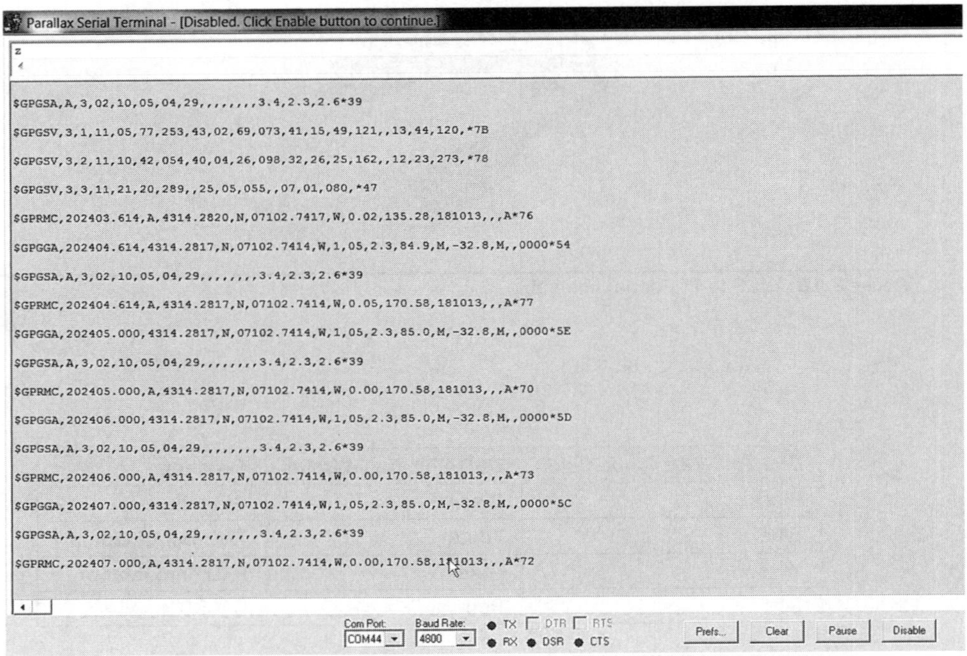

```
Parallax Serial Terminal - [Disabled. Click Enable button to continue.]

z

$GPGSA,A,3,02,10,05,04,29,,,,,,,,3.4,2.3,2.6*39

$GPGSV,3,1,11,05,77,253,43,02,69,073,41,15,49,121,,13,44,120,*7B

$GPGSV,3,2,11,10,42,054,40,04,26,098,32,26,25,162,,12,23,273,*78

$GPGSV,3,3,11,21,20,289,,25,05,055,,07,01,080,*47

$GPRMC,202403.614,A,4314.2820,N,07102.7417,W,0.02,135.28,181013,,,A*76

$GPGGA,202404.614,4314.2817,N,07102.7414,W,1,05,2.3,84.9,M,-32.8,M,,0000*54

$GPGSA,A,3,02,10,05,04,29,,,,,,,,3.4,2.3,2.6*39

$GPRMC,202404.614,A,4314.2817,N,07102.7414,W,0.05,170.58,181013,,,A*77

$GPGGA,202405.000,4314.2817,N,07102.7414,W,1,05,2.3,85.0,M,-32.8,M,,0000*5E

$GPGSA,A,3,02,10,05,04,29,,,,,,,,3.4,2.3,2.6*39

$GPRMC,202405.000,A,4314.2817,N,07102.7414,W,0.00,170.58,181013,,,A*70

$GPGGA,202406.000,4314.2817,N,07102.7414,W,1,05,2.3,85.0,M,-32.8,M,,0000*5D

$GPGSA,A,3,02,10,05,04,29,,,,,,,,3.4,2.3,2.6*39

$GPRMC,202406.000,A,4314.2817,N,07102.7414,W,0.00,170.58,181013,,,A*73

$GPGGA,202407.000,4314.2817,N,07102.7414,W,1,05,2.3,85.0,M,-32.8,M,,0000*5C

$GPGSA,A,3,02,10,05,04,29,,,,,,,,3.4,2.3,2.6*39

$GPRMC,202407.000,A,4314.2817,N,07102.7414,W,0.00,170.58,181013,,,A*72
```

Com Port: COM44   Baud Rate: 4800   TX DTR RTS   RX DSR CTS   Prefs...   Clear   Pause   Disable

FIGURE 8.12   Propeller Serial Terminal screen capture of a GPS data stream.

start using the GPS receiver, but first I will discuss the NMEA protocol and the messages that are being generated from the PMB-688 GPS receiver.

NMEA Protocol

As noted previously, NMEA is the acronym for the *National Marine Electronics Association*, but nobody refers to it by its formal name. NMEA is the originator and continuing sponsor of the NMEA 0183 standard that defines, among other things, the electrical and physical standards to be used in GPS receivers. This standard specifies a series of message types that receivers use to create messages that conform to the following rules, also known as the *Application Layer Protocol Rules*:

- The starting character in each message is the dollar sign.
- The next five characters are composed of the talker id (first two characters) and the message type (last three characters).
- All data fields that follow are delimited by commas.
- Unavailable data is designated by only the delimiting comma.
- The asterisk character immediately follows the last data field, but only if a checksum is applied.
- The checksum is a two-digit hexadecimal number that is calculated using a bitwise exclusive OR algorithm on all the data between the starting '$' character and the ending '*' character as well as including those characters.

There are a large variety of messages available in the NMEA standard; however, the following subset is applicable to the GPS environment and is shown in Table 8.3. All GPS messages start with the letters "GP."

Latitude and Longitude Formats

The two digits immediately to the left of the decimal point are whole minutes; those to the right are decimals of minutes. The remaining digits to the left of the whole minutes are whole degrees.

Message Prefix	Meaning
AAM	Waypoint arrival alarm
ALM	Almanac data
APA	Auto pilot A sentence
APB	Auto pilot B sentence
BOD	Bearing origin to destination
BWC	Bearing using great circle route
DTM	Datum being used
GGA	Fix information
GLL	Lat/Lon data
GRS	GPS range residuals
GSA	Overall satellite data
GST	GPS psuedorange noise statistics
GSV	Detailed satellite data
MSK	Send control for a beacon receiver
MSS	Beacon receiver status information
RMA	Recommended Loran data
RMB	Recommended navigation data for GPS
RMC	Recommended minimum data for GPS
RTE	Route message
TRF	Transit-fix data
STN	Multiple data id
VBW	Dual ground/water speed
VTG	Vector track a speed over the ground
WCV	Waypoint closure velocity (velocity made good)
WPL	Waypoint location information
XTC	Cross-track error
XTE	Measured cross-track error
ZTG	Zulu (UTC) time and time to go (to destination)
ZDA	Date and time

TABLE 8.3 NMEA GPS Message Types

Examples

4224.50 is 42 degrees and 24.50 minutes, or 24 minutes, 30 seconds. The ".50" of a minute is exactly 30 seconds.

7045.80 is 70 degrees and 45.80 minutes, or 45 minutes, 48 seconds. The ".80" of a minute is exactly 48 seconds.

Parsed GPS Message

The following is an example of a parsed GPGLL message that illustrates how to analyze an actual data message:

$GPGLL,5133.80,N,14240.25,W*75
　1 2　　3　　　　4　　5　　　6 7

1	GP	GPS NMEA designator
2	GLL	Lat/Lon message type
3	5133.80	Current latitude 51 degrees, 33 minutes, 48 seconds
4	N	North/South
5	14240.25	Current longitude 142 degrees, 40 minutes, 15 seconds
6	W	East/West
7	*75	checksum

All GPS applications use some type of parser application to analyze data messages and extract the required information to meet system requirements.

Propeller Mini

At this point I would like to take a moment to introduce you to the Propeller Mini module, which I used as the onboard processor for the GPS receiver. The Propeller Mini, which I will henceforth refer to as the Mini, is a relatively new Parallax module, just introduced in 2013. The module is very reasonable in cost and fully supports the full Propeller Spin programming language as well as any other Propeller-compatible language. Figure 8.13 is a picture of the Mini.

Figure 8.13 Parallax Propeller Mini module.

Features/Specifications	Description
Voltage requirements	Regulated 6.5–12 V DC through the VIN pin
Communication	Mini USB, the Propeller Plug required for programming (#32210, not included)
Dimensions	0.81 × 1.52 in (20.5 × 38.6 mm)
Power outputs	3.3 V DC regulated output @ 400 mA max
	5 V DC regulated output @ 600 mA max
Operating temp range	−40°F to +185°F (−40°C to +85°C)
GPIO pins	19, P0 to P18

TABLE 8.4 Key Details and Specifications for the Parallax Propeller Mini Module

Key details and specifications for the Mini are shown in Table 8.4. I mounted the Mini on a solderless breadboard to gain easy access to all the *general-purpose input/output* (GPIO) pins. The breadboard will also allow for easy mounting and connections to both the GPS receiver and the XBee transceiver, which is discussed in the following section.

Radio-Frequency Transceiver Module

GPS data must be sent wirelessly from the quadcopter to a ground station where it is received and displayed for the operator's use. I selected XBee transceivers to perform this function, since they are small, lightweight, inexpensive, and totally compatible with the other modules used in this project. XBee is the brand name for a series of digital RF transceivers manufactured by Digi International. Figure 8.14 shows one of the XBee Series 1 transceivers that I used.

There are two rows of 10 pins on each side of the module. These pins are spaced at 2 mm between each one, which is incompatible with the standard 0.1-in spacing used on solderless breadboards. This means that special connector sockets must be used with the XBee modules. Fortunately, Parallax has anticipated this issue and has provided several solutions.

FIGURE 8.14 XBee Series 1 transceiver.

FIGURE 8.15   Parallax XBee SIP Adapter.

I used two different approaches to mounting the XBee modules. The first was to use Parallax's XBee SIP Adapter, which is shown in Figure 8.15 with an XBee mounted on it. SIP is an abbreviation for *single inline package,* which is a strange description if you look at the adapter's bottom edge where two rows of pins are attached. The two-pin rows are just to provide mechanical stability, since the two pins in each row are electrically connected to one another.

The other mounting approach is actually part of the Parallax's Propeller Board of Education (BOE) development board. The Parallax BOE designers thought that the XBee would be a very popular peripheral for this board, so they incorporated two 10-pin-row mounting sockets. The BOE is shown in Figure 8.16 with the XBee socket visible at the bottom

FIGURE 8.16   The BOE.

center of the board. There is a microSD socket situated between the XBee sockets. The BOE designers also included eight socket pins where key XBee pins are easily available for interconnections. These pins are part of a 10×2 socket located above the right XBee socket.

I will next examine the XBee hardware to show how this clever design makes wireless transmission so easy.

XBee Hardware

All the electronics in the XBee hardware, except for the antenna, are contained in a slim metal case located on the bottom side of the module, as may be seen in Figure 8.17. If you look closely at the figure, you should see the bottom of the antenna wire, which is located near the top left corner of the case. While Digi International is not too forthcoming regarding what makes up the electronic contents of the case, I did determine that the earlier versions of the Series 1 XBee transceivers used the Freescale™ model MC13192 RF transceiver. This chip is a hybrid type, meaning that it is made up of both analog and digital components. The analog components make up the RF transmit-and-receive circuits while the digital components implement all the other chip functions. It is a complex chip, which is the reason why the XBee module is so versatile and able to automatically perform a remarkable number of networking functions. Table 8.5 shows a select number of features and specifications for the MC13192.

The XBee module implements a full network protocol suite (which is discussed below in the software section), but from a hardware perspective, it means that there must also be a microprocessor present in the electronics case. From my research, I cannot determine which type of microprocessor it is, but I am willing to make an educated guess that it would be a Freescale™ chip, based upon the reasonable assumption that the MC13192 would be designed to be highly compatible with the company's own line of microprocessors. One other factor supporting my guess is that Digi International has recently introduced a line of programmable XBee modules named XBee Pro SB that use the 8-bit Freescale™ S08 microprocessor. Of course, being able to put your own programs into the XBee would eliminate the need for the Mini, but that would not be as much fun and would probably be a bit limiting, given the tremendous capabilities of the Propeller chip.

Figure 8.17 XBee electronics case.

Features/Specifications	Description
Frequency/modulation	O-QPSK data in 5.0 MHz channels and full spread-spectrum encode and decode (modified DSSS) Operates on one of 16 selectable channels in the 2.4 GHz ISM band
Maximum bandwidth	250 kbps (compatible with the 802.15.4 Standard)
Receiver sensitivity	<-92 dBm (typical) at 1.0% packet error rate
Maximum output power	0 dBm nominal, programmable from -27 dBm to 4 dBm
Power supply	2.0 to 3.4 V
Power conservation modes	< 1 μA Off current 1 μA Typical hibernate current 35 μA Typical doze current (no CLKO)
Timers/comparators	Four internal timer comparators available to supplement MCU resource
Clock outputs	Programmable frequency clock output (CLKO) for use by MCU
Number of GPIO pins	7
Internal oscillator	16 MHz with onboard trim capability
Operating temperature range	-40 to 85°C
Package size	QFN-32 Small form factor (SFF)

TABLE 8.5   Freescale™ MC13192 Features and Specifications

The XBee pins are detailed in a logical arrangement in Figure 8.18 for your information. Just be aware that only four of the pins are needed for this project, and they are shown with an asterisk next to the pin label. All the pin and function descriptions are shown in Table 8.6.

There are a considerable number of functions available to you if needed; however, this project requires only the most minimal functions for simple and reliable data transfers. Thankfully, the XBees automatically connect and establish reliable communications.

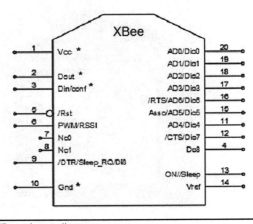

FIGURE 8.18   Logical XBee pinout diagram.

Pin Number	Name(s)	Description
1	Vcc	Power supply, 3.3 V
2	Dout	Data out (TXD)
3	Din	Data in (RXD)
4	DIO12	GPIO pin 12
5	Reset	XBee module reset, pin low
6	PWM0/RSSI/DIO10	*Pulse width modulation* (PWM) Analog 0, *received signal strength indicator* (RSSI), GPIO pin 10
7	DIO7	GPIO pin 7
8	Reserved	*Do Not Connect* (DNC)
9	DTR/SLEEP_RQ/DIO8	*Data Terminal Ready* (DTR), GPIO Sleep Assertion (pin low), GPIO pin 8
10	GND	Ground or common
11	DIO4	GPIO pin 4
12	CTS/DIO7	*Clear To Send* (CTS), GPIO pin 7
13	ON/SLEEP	Pin high when NOT sleeping
14	Vref	Voltage reference level (used with analog-to-digital conversion)
15	ASSOC/DIO5	Pulse signal when connected to a network, GPIO pin 5
16	RTS/DIO6	*Request To Send* (RTS), GPIO pin 6
17	AD3/DIO3	Analog Input 3, GPIO pin 3
18	AD2/DIO2	Analog Input 2, GPIO pin 2
19	AD1/DIO1	Analog Input 1, GPIO pin 1
20	AD0/DIO0/COMMIS	Analog Input 0, GPIO pin 0, Commissioning Button

TABLE 8.6   XBee Pin Descriptions and Functions

What is even more interesting than the hardware design is the data transmission and reception protocol that XBee implements, which I will discuss in the next section.

XBee Data Protocol

The XBee uses a highly capable networking protocol named ZigBee, which is also called a *Personal Area Network* (PAN). I will endeavor to keep the technical jargon to a minimum; however, it is important that you get at least a fundamental knowledge of how the ZigBee network functions in case something does not work as planned.

Zigbee was designed to be compliant with the ISO 7 Layer network model. As such, its inherent design is based upon proven computer network concepts that are robust, efficient, and well understood by most system designers. Figure 8.19 shows the ZigBee logical network stack with the corresponding ISO layer number. All subsequent network software developed for the ZigBee network follows this model.

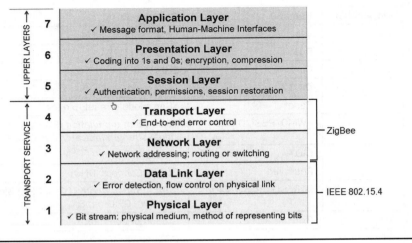

FIGURE 8.19   ZigBee and the ISO network layers.

Data sent through the ZigBee network is in packets similar to the Ethernet format. Figure 8.20 shows how these packets are initially constituted at Layer 2, or MAC as it is referred to in the figure. These packets may be subsequently modified at higher layers, as needed, to suit the real-time network communication needs.

There are four packet types that exist in ZigBee:

1. Beacon
2. Data
3. MAC Command
4. ACK

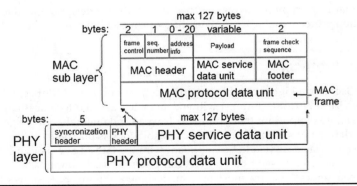

FIGURE 8.20   ZigBee packet formation.

Actual data packets are formed at the MAC, or layer 2, level where the data is prepended with both the source and destination addresses. A sequence number is also assigned to allow the receiver to determine the correct sequence of received packets. It is relatively easy to receive out-of-sequence packets in this type of network. Frame control bytes are also appended for error checking, which is the reason why ACK packets are required. ZigBee is a type of connection network, similar to Ethernet, that has a very robust way of ensuring that packets get where they need to go. ZigBee Layer 3 uses *acknowledgement packet (ACK)*.

The receiver performs a 16-bit *cyclic redundancy check* (CRC) to verify that the packet was not corrupted during transmission. If a good CRC is determined, the receiver will then transmit an ACK; this action allows the transmitting XBee node to know that the data were received properly. The packet is discarded if the CRC indicates the packet was corrupted, and no ACK is transmitted. The network should be configured such that the transmitting node will resend up to a predetermined number of times until either the packet is successfully received or the resend limit is reached. The ZigBee protocol provides self-healing capabilities if the path between the transmitter and receiver has become unreliable or a complete network failure has happened. Alternate paths will be established if physically possible.

Layers 1 and 2 support the following standards:

- Star, mesh, and cluster tree topologies
- Beaconed networks
- GTS for low latency
- Multiple power-saving modes (idle, doze, hibernate)

Layers 3 and 4 further refine the packets by identifying what the packet type is, where it is going, and where it has been. They also set the data payload and support the following:

- Point-to-point and star network configurations
- Proprietary networks

Layer 4 sets up the routing, thus ensuring that the packets are sent along the correct paths to reach the desired nodes. This layer also ensures that:

- ZigBee 1.0 specifications are met
- Support is provided for star, mesh, and tree networks

There are also three ZigBee standards that primarily involve Layers 3 and 4. These standards are:

1. Routing—Defines how messages are sent through ZigBee nodes. Also referred to as *digi-peating*.
2. Ad hoc network—Creates a network automatically without any operator involvement.
3. Self-healing mesh—Determines automatically if a malfunctioning node exists and reroutes messages, if physically possible.

Layer 5 is responsible for security, which is enforced by using the *Advanced Encryption Standard* (AES) 128-bit security key.

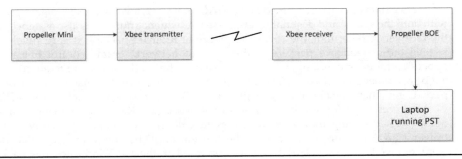

FIGURE 8.21   Diagram of XBee functional test.

XBee Functional Test

It is now time to demonstrate how the XBee works by using a simple test configuration between two XBee nodes and the Propeller boards controlling these nodes. Figure 8.21 shows the test configuration in which the XBee transmitter is being controlled by the Mini, and the XBee receiver is being controlled by the BOE.

Two separate programs need to be loaded into each of the Propeller boards, one for transmit and the other for receive. The transmit program is named Test XBee Transmit.spin, and the source code is shown below:

```
OBJ
  system : "Propeller Board of Education"  ' PropBOE configuration
                                             tools
  time     : "Timing"                      ' Timing convenience
                                             methods
  xb       : "XBee_Object_1"               ' XBee communication
                                             methods


PUB Start
  system.Clock(80_000_000)                 ' System clock -> 80 MHz
  xb.start(7, 6, 0, 9600)                  ' Propeller Comms - RX,
                                             TX,  Mode, Baud

  xb.AT_Init                               ' Initialize for fast AT
                                             command use.

  xb.AT_ConfigVal(string("ATMY"), 8)       ' Set MY address to 8
  xb.AT_ConfigVal(String("ATDL"), 9)       ' Set destination low
                                             address to 9

  repeat
    xb.str(string("This is a test. ", 13)) ' Send a string
    time.pause(500)                        ' Wait half a second
```

This program and the following program were downloaded from the Parallax OBEX forum. Both programs were very slightly modified for this project. Note that the transmit

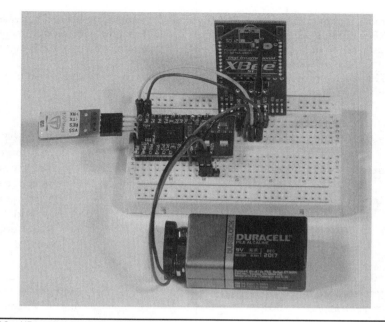

FIGURE 8.22   XBee and Propeller Mini transmitter node.

program uses a BOE object named *Propeller Board of Education* and referenced as *system*, which also works perfectly with the Mini board. I am constantly impressed with how well Propeller software objects function among different development environments. It is a testament to the simplicity and consistent architecture used in the Parallax programming languages.

Figure 8.22 is a photo of the XBee mounted on an SIP adapter that is connected to the Mini—all mounted on a solderless breadboard. The whole transmitter assembly is powered by a single 9-V battery, which is also shown in the photo.

The Propeller Plug programming tool is also shown attached to the Mini in the figure. It is needed only to program the Mini for this project. There are four connections needed between the Mini and the XBee module, which are shown in Table 8.7. Of course, the Mini must be powered, which in this case, is with a 9-V battery connected to VIN and GND. Be sure you watch the polarity connection.

The transmit program will continuously send the phrase, *This is a test,* two times per second. The complimentary receiver node is composed of an XBee mounted on a BOE. This

XBee Module	Propeller Mini
+5 V	5 V
GND	GND
DOUT	P7
DIN	P6

TABLE 8.7   XBee and Propeller Mini connections.

assembly is shown in Figure 8.23. Only two connections are needed between the BOE and the XBee module, as shown in Table 8.8. The BOE is powered through the mini USB cable that attaches to the laptop running the PSerT program, while the XBee module is powered through the BOE sockets.

The receive program is named Test XBee Receive.spin, and the source code is shown below:

```
OBJ
  system : "Propeller Board of Education"      ' PropBOE
                                                 configuration tools
  pst        : "Parallax Serial Terminal Plus"  ' Terminal
                                                 communication tools
  time       : "Timing"                         ' Timing convenience
                                                 methods
  xb         : "XBee_Object_1"                   ' XBee communication
                                                 methods


PUB Go | c
  system.Clock(80_000_000)                      ' System clock -> 80 MHz
  pst.Start(115_200)                            ' Start Parallax Serial
                                                 Terminal

  xb.start(7,6,0,9600)                          ' Propeller Comms -
                                                 RX,TX, Mode, Baud

  xb.AT_Init                                    ' Initialize for fast AT
                                                 command use.
```

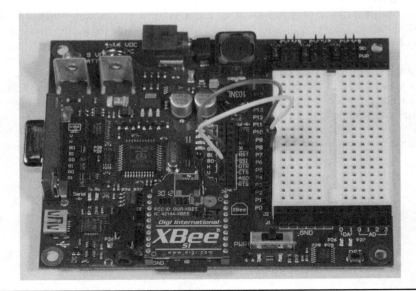

Figure 8.23 XBee and Propeller BOE receiver node.

XBee Module	Propeller BOE
DO	P7
DI	P6

TABLE 8.8   XBee and Propeller BOE Connections

```
xb.AT_ConfigVal(string("ATMY"), 9)        ' Set MY address to 9
xb.AT_ConfigVal(String("ATDL"), 8)        ' Set Destination Low
                                            address to 8

repeat                                     ' Main loop
  c := xb.rxCheck                          ' Check buffer
  if c <> -1                               ' If it's not empty (-1)
    pst.Char(c)                            ' Then display the
                                            character
```

The receive program uses the same objects as the transmit program with the addition of the Parallax Serial Terminal Plus object, which enables the display of the received data on the laptop screen that is running the PSerT program. Figure 8.24 is a screenshot of the laptop screen that is running the PSerT application for the data transmission test.

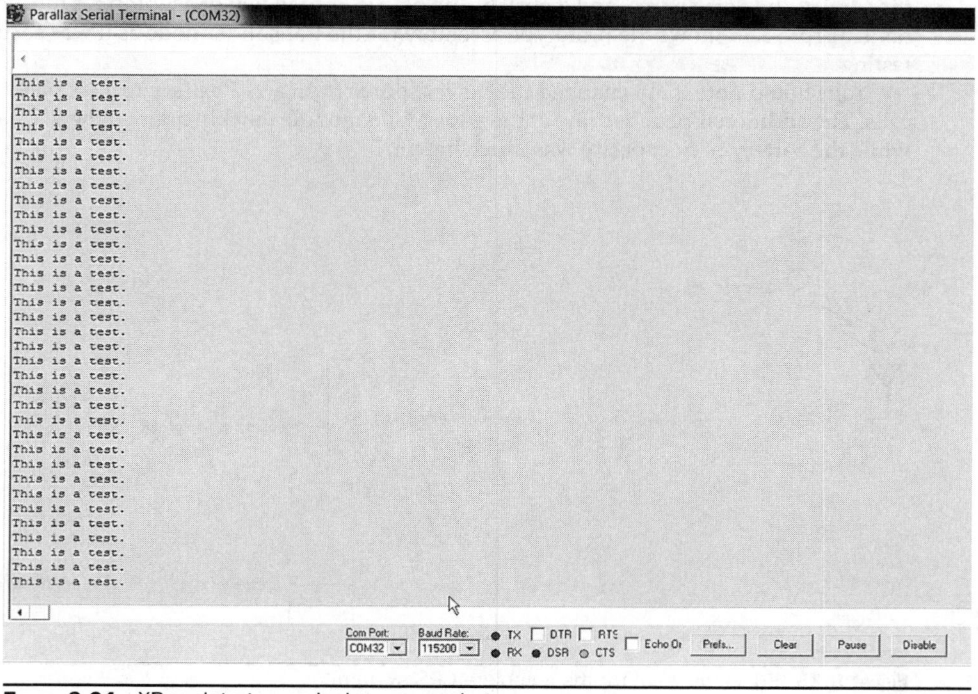

FIGURE 8.24   XBee data transmission screenshot.

XBee Range Check

I performed a simple range check to determine the approximate operating range for the XBee test system. The transmitter node was set on a tripod at one end of a large, open field. I then walked away from the transmitter while carrying a laptop running the test program. I walked approximately 114 paces from the transmitter, at which point the transmission became intermittent. My pace is about one meter, so that was a good estimate for the reliable range. I walked a bit further and continually repositioned the receiver node to see if the signal could be reacquired. I was able to go out to about 154 m, at which point no amount of node juggling could get the signal back. The 114-m distance is actually a bit better than the stated 100-m range in the Zigbee specification.

Digi International also manufactures a line they call the XBee Pro, which generates up to 60 mW of power—much greater than the regular Series 1 power output of 1 mW. The Pro brochure claims a line-of-sight range of up to one mile. That is an impressive range, but I am fairly sure that it would far exceed the range of the R/C transmitter and the FPV camera system, which will be discussed in the next chapter. In any case, I believe that the quadcopter would be invisible to the operator at such a distance, which is never a good idea.

The range check confirmed that the XBee modules appropriately support operations at or below 300 feet when they are above ground and reasonably close to the operator's control station.

Complete GPS System

A block diagram of the complete GPS system is shown in Figure 8.25.

In the complete system, unlike the test system, the PMB-688 GPS module is connected to the Mini as the data source, and a portable display is used in lieu of a laptop for the receiver node. Figure 8.26 shows the prototype assembly for the transmitter node as it was set up for testing.

You should note that I changed the power source from a 9-V battery to a six pack of AA cells. The additional current draw of the added GPS module quickly drained the 9-V battery, while the battery pack capacity was much higher.

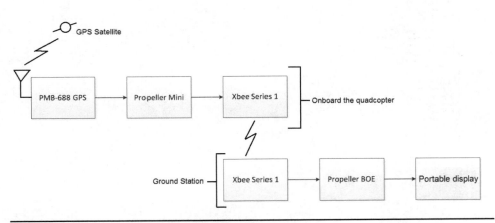

Figure 8.25   Block diagram for the complete GPS system.

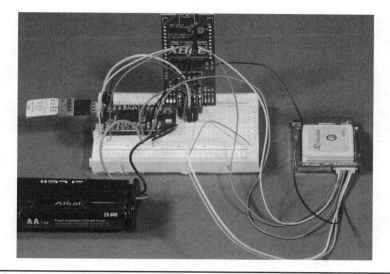

FIGURE 8.26   Prototype GPS transmitter node.

All GPS and Mini interconnections are shown in Table 8.9. All of the interconnections between the XBee and Mini are still in place and remain as they were detailed in Table 8.7.

Note that the GPS yellow wire is named TTLTX, which means that data are output via this wire. This yellow wire is actually the receive line, which can be a bit confusing. Just remember that data-communication port nomenclature is usually specified from the perspective of the module, that is, data coming out of the module is TX, while data going into the module is RX.

The program that is run in the Mini is significantly different from the test program shown above. It now incorporates a GPS driver object as well as the existing XBee object. It also contains a considerable amount of code to parse, or separate, the raw NMEA data that

GPS Module	Propeller Mini
+5 V (Red)	5 V
GND (Black)	GND
TTLTX (Yellow)	P8
TTLRX (Blue)	P9

TABLE 8.9   GPS and Propeller Mini Connections

is streaming from the GPS PMB-688 module. The Spin object is named GPS_Propeller, and the code is shown below.

```
{{
GPS_XBee
Modified !GPS_Propeller program by D. J. Norris 2013
This program controls an Elev-8 Quadcopter real-time data system
Pin connections:
  Transmitter:
      XBee DOUT        to P7
      XBee DIN         to P6
      GPS Data Out     to P8
      GPS Data In      to P9
  Receiver:
      XBee DOUT        to P7
      XBee DIN         to P6
}}

CON
  _clkmode = xtal1 + pll16x
  _xinfreq = 5_000_000

  XB_Rx        = 7            ' XBee DOUT
  XB_Tx        = 6            ' XBee DIN
  XB_Baud      = 9600
  CR           = 13          ' Carriage Return value
  GPS_Pin      = 8

OBJ
   GPS        : "GPS_Float_Lite"
   FS         : "FloatString"
   xb1        : "XBee_Object_1"

Pub  Start | fv
  xb1.start(XB_Rx, XB_Tx, 0, XB_Baud) ' Initialize comms for XBee
  xb1.AT_Init
  xb1.AT_ConfigVal(string("ATMY"),8)  ' My XBee address is set to 8
  xb1.AT_ConfigVal(string("ATDL"),9)  ' The remote XBee address is
                                        set to 9
  GPS.Init

  repeat
    xb1.Str(String(16, 1))

    FS.SetPrecision(7)
    fv := GPS.Float_Latitude_Deg       ' Get latitude
```

```
  If fv <> floatNaN
    xb1.Str(FS.FloatToString(fv))
  Else
    xb1.Str(String("---"))


  xb1.Str(String(","))
  fv := GPS.Float_Longitude_Deg          ' Get longitude
  If fv <> floatNaN
    xb1.Str(FS.FloatToString(fv))
  Else
    xb1.Str(String("---"))


  xb1.Str(String(","))
  fv := GPS.Float_Speed_Over_Ground      ' Get speed
  If fv <> floatNaN
    xb1.Str(FS.FloatToString(fv))
  Else
    xb1.Str(String("---"))


  xb1.Str(String(","))
  fv := GPS.Float_Altitude_Above_MSL     ' Get altitude
  If fv <> floatNaN
    xb1.Str(FS.FloatToString(fv))
  Else
    xb1.Str(String("---"))


  xb1.Str(String(","))
  fv := GPS.Long_Month                   ' Get month
  If fv <> floatNaN
    xb1.Dec(GPS.Long_Month)
  Else
    xb1.Str(String("---"))


  xb1.Str(String(","))
  fv := GPS.Long_Day                     ' Get day
  If fv <> floatNaN
    xb1.Dec(GPS.Long_Day)
  Else
    xb1.Str(String("---"))


  xb1.Str(String(","))
  fv := GPS.Long_Year                    ' Get year
  If fv <> floatNaN
    xb1.Dec(GPS.Long_Year)
  Else
    xb1.Str(String("---"))
```

```
        xb1.Str(String(","))
        fv := GPS.Long_Hour                      ' Get Hour
        If fv <> floatNaN
          xb1.Dec(GPS.Long_Hour)
        Else
          xb1.Str(String("---"))

        xb1.Str(String(","))
        fv := GPS.Long_Minute                    ' Get Minute
        If fv <> floatNaN
          xb1.Dec(GPS.Long_Minute)
        Else
          xb1.Str(String("---"))

        xb1.Str(String(","))
        fv := GPS.Long_Second                    ' Get Second
        If fv <> floatNaN
          xb1.Dec(GPS.Long_Second)
        Else
          xb1.Str(String("---"))
        xb1.tx(13)
        WaitCnt(ClkFreq / 2 + ClkFreq / 4 + Cnt) ' Wait .75 seconds to
repeat


DAT
     floatNaN          LONG $7FFF_FFFF           'Means Not a Number
```

The above program uses the same XBee driver object, XBee_Object_1, as was the case in the previous test programs. This program also uses a GPS driver object named GPS_Float_Lite that takes care of all the necessary protocols between the GPS module and the Mini. Retrieving data from the GPS module becomes very easy, since all that is required is to call the GPS driver method that returns the desired data. For example, the following gets the current GPS hour data:

```
fv := GPS.Long_Hour
```

Where fv is a local variable, GPS is the local reference to GPS_Float_Lite, and Long_Hour is the name of the method in GPS_Float_Lite that returns the current GPS UTC hour value. UTC is short for *Coordinated Universal Time* and is the time standard used to ensure that GPS time reports become independent of any time zone.

Except for the deployed display, the receive portion of this system is the same as the one previously described in the test system. I used the BOE with the laptop to test the prototype system, which made sense because the portable display would not impact any essential data transfers between the transmitter and receiver XBee nodes. The receiver-node connections were detailed in Table 8.8, and the assembly was shown in Figure 8.23. Figure 8.27 is a laptop screenshot showing the results of a trial run with the complete GPS system. The 10 data elements displayed in the figure are detailed in Table 8.10 for your information. GPS data is

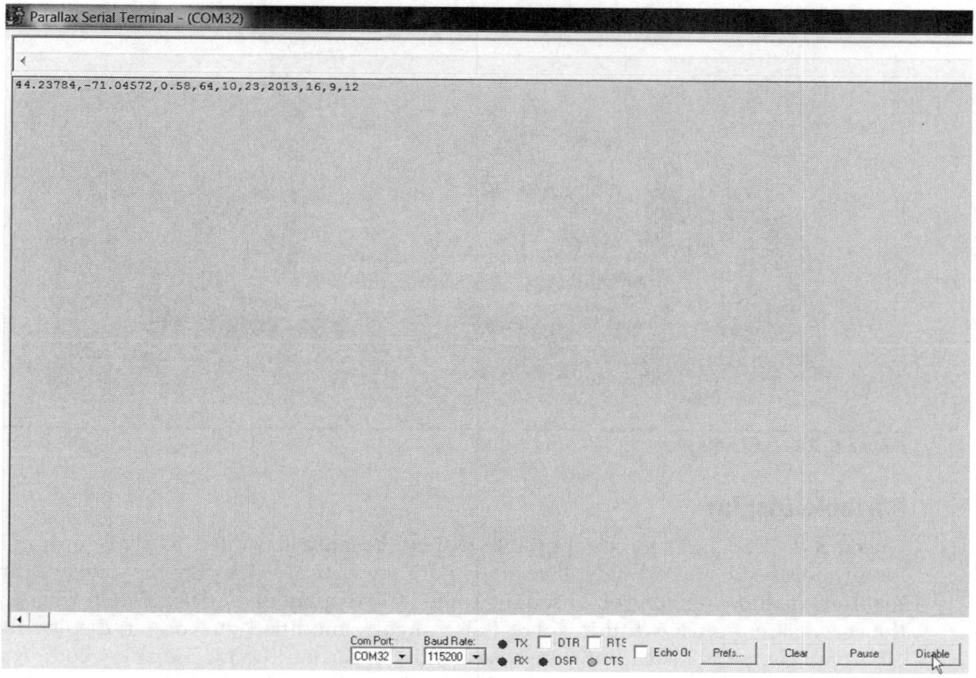

FIGURE 8.27   GPS data displayed by the PSerT application.

updated every 0.75 seconds, which allows sufficient time for the GPS module to receive a fresh data set. Please note that the GPS module provides the coordinates in *decimal degrees* with the minutes and seconds as a decimal fraction of a degree in lieu of the NMEA format discussed earlier. It is fairly easy to make the conversion to integer minutes and seconds using software, if that is the required format.

Element #	Value	Description
1	44.23784	Latitude (North)
2	−71.04572	Longitude (West)
3	0.58	Speed (m/s)
4	64	Altitude (m)
5	10	Month
6	23	Day
7	2013	Year
8	16	Hours (UTC)
9	9	Minutes
10	12	Seconds

TABLE 8.10   Displayed GPS data elements.

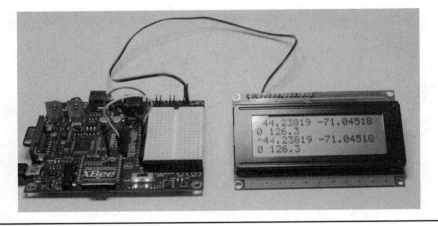

FIGURE 8.28 LCD display.

Portable Display

I used a 4 × 20 LCD for the portable display because I wanted to show only latitude, longitude, speed, and altitude. It really is not very critical to display the date and time for real-time quadcopter control. I used the same LCD display that I described in Chapter 7 for the servo-signal pulse-width display. I also slightly modified the code to display the four GPS data elements instead of the pulse widths. Inserting the LCD display code was also quite easy, since I only had to add the LCD display-driver object and reference the input data pin, which is P19. Incidentally, I also left the PSerT driver code in place, since I figured it might come in handy for any debugging.

Figure 8.28 shows the LCD display with the four GPS data elements. They are repeated twice because of the simple character-by-character transmission scheme that is being used. I do not consider this side effect to be much of an issue.

The modified Test XBee Receive program is shown below.

```
{{
Test XBee Receive program modified to display both on PST and LCD
screens
D. J. Norris  2013
}}

CON
  _clkmode = xtal1 + pll16x
  _xinfreq = 5_000_000

  LCD_PIN   = 19
  LCD_BAUD  = 19_200
  LCD_LINES = 4

OBJ
  system : "Propeller Board of Education"    ' PropBOE configuration
                                               tools
```

```
pst    : "Parallax Serial Terminal Plus"   ' Terminal
                                             communication tools
time   : "Timing"                          ' Timing convenience
                                             methods
xb     : "XBee_Object_1"                    ' XBee communication
                                             methods
lcd    : "debug_lcd"


PUB Go | c

  pst.Start(115_200)                        ' Start Parallax
                                              Serial Terminal

  xb.start(7,6,0,9600)                      ' Propeller Comms -
                                              RX, TX, Mode, Baud

  xb.AT_Init                                ' Initialize for fast
                                              AT command
  xb.AT_ConfigVal(string("ATMY"), 9)        ' Set MY address to 9
  xb.AT_ConfigVal(String("ATDL"), 8)        ' Set Destination Low
                                              address to 8


  if lcd.init(LCD_PIN, LCD_BAUD, LCD_LINES) ' init returns True
                                              if started
    lcd.cursor(0)                           ' cursor off
    lcd.backLight(True)                     ' turn on backlight
    lcd.cls                                 ' clear display
    lcd.str(string("GPS Real Time Display V1.0"))  ' welcome
                                                     screen
    waitcnt(clkfreq + cnt)                  ' wait 1 second
    lcd.cls                                 ' clear the screen
                                              again
    waitcnt(clkfreq/2 + cnt)                ' wait .5 seconds


  repeat                                    ' Main forever loop
    c := xb.rxCheck                         ' Check buffer
    if c <> -1                              ' If it's not
                                              empty (-1)

      pst.Char(c)                           ' Then display the
                                              character on the
                                              serial terminal

      lcd.putc(c)                           ' Display the same
                                              character on the
                                              LCD screen
```

I chose to connect the LCD display to pin 19 because that is also a BOE servo port. That way, I could use a servo cable to connect the LCD display with the BOE as I did in Chapter 7.

The transmit code was modified to transmit only the four GPS data elements discussed above. I also eliminated the commas between the data elements to conserve space on the LCD display. The modified transmit program named GPS_XBee_Brief is shown below.

```
{{
GPS_XBee_Brief
Modified !GPS_Propeller program by D. J. Norris 2013
This program controls an Elev-8 Quadcopter real-time data system
This brief edition transmits only four data elements:
   Latitude
   Longitude
   Speed
   Altitude

Pin connections:
   Transmitter:
      XBee DOUT to P7
      XBee DIN  to P6
      GPS Rx    to P8
      GPS Tx    to P9
   Receiver:
      XBee DOUT to P7
      XBee DIN  to P6
}}
CON
  _clkmode    = xtal1 + pll16x
  _xinfreq    = 5_000_000

  XB_Rx       = 7      ' XBee DOUT
  XB_Tx       = 6      ' XBee DIN
  XB_Baud     = 9600
  CR          = 13     ' Carriage Return value
  GPS_Pin     = 8
OBJ
   GPS        : "GPS_Float_Lite"
   FS         : "FloatString"
   xb1        : "XBee_Object_1"

Pub  Start | fv
xb1.start(XB_Rx, XB_Tx, 0, XB_Baud)                ' Initialize comms for
                                                     XBee

xb1.AT_Init
xb1.AT_ConfigVal(string("ATMY"),8)
xb1.AT_ConfigVal(string("ATDL"),9)
GPS.Init
```

```
repeat
    xbl.Str(String(16, 1))

    FS.SetPrecision(7)
    fv := GPS.Float_Latitude_Deg              ' Get latitude
    If fv <> floatNaN
      xbl.Str(FS.FloatToString(fv))
    Else
      xbl.Str(String("---"))

    xbl.Str(String(" "))
    fv := GPS.Float_Longitude_Deg             ' Get longitude
    If fv <> floatNaN
      xbl.Str(FS.FloatToString(fv))
    Else
      xbl.Str(String("---"))

    xbl.Str(String(" "))
    fv := GPS.Float_Speed_Over_Ground         ' Get speed
    If fv <> floatNaN
      xbl.Str(FS.FloatToString(fv))
    Else
      xbl.Str(String("---"))

    xbl.Str(String(" "))
    fv := GPS.Float_Altitude_Above_MSL        ' Get altitude
    If fv <> floatNaN
      xbl.Str(FS.FloatToString(fv))
    Else
      xbl.Str(String("---"))

    xbl.tx(13)
    WaitCnt(ClkFreq / 2 + ClkFreq / 4 + Cnt)

DAT
    floatNaN        LONG $7FFF_FFFF           ' Means Not a Number
```

Mounting the Transmitter XBee Node

Figure 8.29 shows the front of the XBee transmitter node assembly. I wanted to use the XBee SIP-mounting adapter but did not want to mount it vertically in a solderless breadboard. I was concerned about the quadcopter vibrations shaking loose the somewhat top-heavy XBee assembly. After a bit of thought, I came up with the configuration that you see in the figure, where the SIP assembly is still plugged into a breadboard. However, the breadboard itself is attached by foam-backed, double-sided tape to another breadboard that holds the Mini. I also put a hidden spacer between the SIP adapter and the horizontal breadboard to

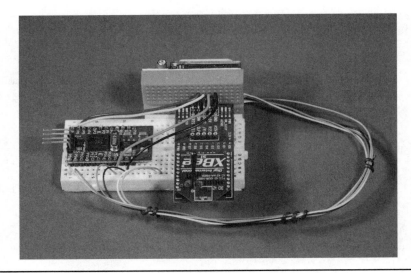

FIGURE 8.29   Front of the XBee transmitter node.

increase the overall rigidity of the assembly. The GPS module is mounted to the back of the vertical breadboard, as you can see in Figure 8.30.

The GPS is held in place by the double-sided tape that is normally attached to the bottom of new breadboards. You should be aware that a vertically mounted GPS module with an internal antenna is not as sensitive to satellite signals as a horizontally mounted one is. You might want to initially hold the quadcopter on its side with the flat side of the GPS antenna pointed to the sky until a signal is acquired. Once the receiver locks on to the satellite signal,

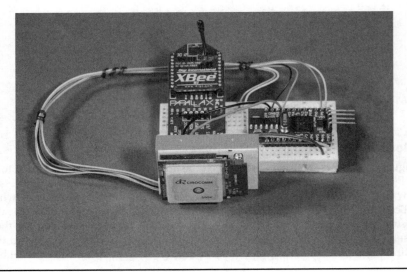

FIGURE 8.30   The back of the XBee transmitter node.

the red LED will be steady instead of blinking. The receiver's hold sensitivity is much better than the acquisition sensitivity, so the quadcopter can be put back to a horizontal position while still locked onto the GPS signals.

Moving Map System

NOTE: *The original intent of this section was to demonstrate how to display the quadcopter's position in real time by using the Google Earth application. Unfortunately, I was never able to stream the raw GPS data successfully from the XBee transmitter to the ground-station XBee receiver, and then into a laptop running Google Earth. However, I do show you how to manually enter the coordinates as they are displayed on the LCD screen so that you can have a near real-time location service.*

I selected the Google Earth application for the moving map display because it incorporates a very convenient interface that accepts serial GPS coordinates and can display them on a computer screen in real time. This project was divided into several phases so that I could experiment with the various technologies involved with the moving map and determine how to best implement each phase. The first phase was simply to use an existing hand-held GPS device plugged into a laptop running the Google Earth application.

Monitoring the Quadcopter Position with the Google Earth Application

Figure 8.31 shows the Google Earth opening screen. I ran the application on a Win7, 11.3-in, and 32-bit Toshiba laptop. You next have to click on the Tools menu bar selection to access the GPS import function. Figure 8.32 shows the real-time GPS menu selection.

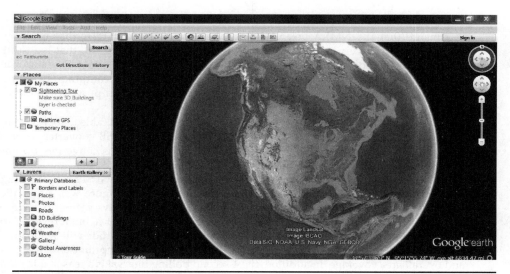

FIGURE 8.31   Google Earth opening screen.

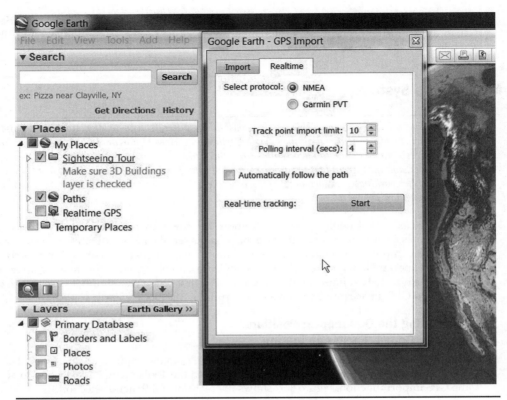

Figure 8.32   Real-time GPS menu screen.

Ensure that the NMEA box is checked because it is the data format used by the majority of GPS receivers. Google Earth also has a very nice feature that scans all the available serial ports in order to identify the active one with the serial GPS data. You can observe this behavior simply by plugging a GPS receiver into a computer's USB port. I used an older style GPS receiver that is shown in Figure 8.33 to provide the GPS data stream.

You can also see in the figure a portion of the older-style RS-232 cable that plugs into the back of the receiver. This type of cable requires an RS-232-to-USB converter to make it compatible with a modern USB port. I used one of the ubiquitous converters commonly available at local office supply stores to adapt my old-style cable.

Figure 8.34 shows the results of a real-time import using the GPS12 receiver plugged into the laptop. The bulls-eye Position indicator shown in the figure was automatically added by the application. Also, the 8000-ft altitude for the view seems to have been automatically selected by the Google Earth program. I tried to lower it, but it reverted to the higher altitude, as shown in the figure. I don't believe this view altitude will be too much of an issue because the 8000-ft, above-ground altitude is a good compromise between detail and overall area coverage. In addition, the 4-second polling interval is a reasonable number, since the Elev-8 should not travel too far in a 4-second period.

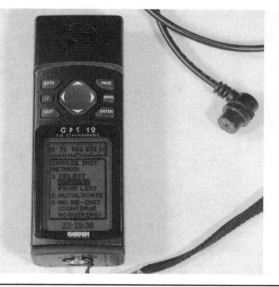

FIGURE 8.33   Garmin GPS12 receiver.

Manual Entry of Position Coordinates into Google Earth

Entering coordinate data into Google Earth is very easy. Simply type the coordinates into the Search text box, and press the Enter key or click on the Search button. For example, I entered the following coordinates, using the decimal-degree format, into the Search text box:

$$42.23978 \text{ N } 71.04598 \text{ W}$$

FIGURE 8.34   Real-time position screenshot.

FIGURE 8.35   Google Earth screenshot.

Ensure that there is one space between each element and that descriptors N and W are also present. Of course, the N and W descriptors will change depending upon where you operate the quadcopter. Figure 8.35 shows the resultant Google Earth screen when the above coordinates were entered into the Search text box.

This section completes my discussion of how to achieve good situational awareness when operating the quadcopter while using a real-time GPS system. The next chapter shows you how to add real-time video capability that will greatly complement this GPS system.

Summary

I began the chapter with a brief history of the GPS system followed by a tutorial that used a fictitious example to explain the basic underlying principles governing the system.

The PMB-688 GPS receiver was discussed next, focusing on its excellent receiver characteristics as well as its easy serial communication link.

I discussed how to set up and test the GPS module with a serial-console link by using a USB/TTL cable. The Propeller Serial Terminal (PSerT) was run on a Windows laptop to verify proper operation of the GPS module.

The NMEA 0183 protocol was thoroughly examined to illustrate the rich set of messages that are created by the GPS receiver. This project uses only a small subset of the data, but you should be aware of what is available for potential use. A look at a parsed GPS message was also included along with a brief explanation of how to interpret latitude and longitude data.

The next item introduced was the new Propeller Mini, which is a very compact Parallax Propeller development module. This module supports the full Spin language with a full complement of *general-purpose input/output* (GPIO) pins. I used the Mini as an onboard controller for both the GPS module and an XBee RF transceiver.

I also carefully explained Zigbee, which is the XBee data communication protocol. It is important to understand how Zigbee works because it is a key part of the control program that runs in the controlling microcontroller.

The XBee functional test demonstrated how simple it is to set up a communications link that could transfer GPS data. The operational software for both the transmitter and receiver nodes was described, and I determined the operational range for the XBee link, which is compatible with the Elev-8 operating profile.

Next, we looked at a complete GPS system and the modified software for the transmitter node. I showed how the system displays GPS data on a PSerT screen and on a portable LCD display. The latter display makes it easy to use in a field-deployable ground station.

I showed a compact design for an onboard GPS/XBee module that may be easily mounted on an Elev-8. The design is resistant to disruptions that might be caused by quadcopter vibrations.

The chapter concluded with a look at how to use the Google Earth application with real-time GPS. I explained how easy it is to enter GPS coordinates manually into Google Earth to show the quadcopter's position in the application.

CHAPTER 9

Airborne Video Systems

Introduction

In this chapter I will discuss two video systems that could easily be carried aloft by an Elev-8 or other quadcopter outfitted with a similar lifting capacity. The first system can record video and stream it back to the operator. It is often referred to as a *first-person video* (FPV) because the operator can use it to view where the quadcopter is going in real time. The FPV used in this chapter is mounted on the tiltable platform described in Chapter 7. By tilting the camera downward, toward the ground, it is possible to observe the ground while hovering. This capability is highly useful for security purposes as well as for activities, such as search and rescue, damage assessment, or wildlife monitoring.

The second video system is a much less expensive one that also transmits real-time video, although the quality of the video is much lower than that of the first system. It also does not have any video-recording capability. However, it is well suited to providing a video source to a post-processing software suite that will be demonstrated later in the chapter. I will refer to this second system as the *economy system* when I discuss it later.

GoPro Hero 3 Camera System

I selected the GoPro Hero 3 Silver edition as the video camera to use in the first system. The Hero 3 line of video cameras is highly popular for use in this type of application as well as in many others, as the large number of YouTube videos produced by this camera system will attest to. I selected the Silver edition as a compromise between cost and features. Table 9.1 is a comparison of the features present in the three currently available versions of the Hero 3 camera.

The Black version is the most feature-packed, and judging from my review of Web blogs, seems to be the most popular. However, the Silver edition was more than adequate for this airborne video application, and the extra money I saved was put to use in other project areas.

The camera is shown outside of its protective case in Figure 9.1. It is a remarkably simple camera with just a few controls. The designers realized that most users would not adjust the camera while they were using it, unlike the users of a *digital single-lens reflex* (DSLR) who constantly adjust or fiddle with the camera. Figure 9.2 is a diagram of the camera front that contains most of the controls.

Model	Pixels (MP)	Photo Burst	Ultra Wide	Wide	Medium	Narrow	Protune
White	5	3	x		x		
Silver	11	10		x	x	x	x
Black	12	30	x	x	x	x	x

TABLE 9.1 Comparison of Hero 3 Camera Versions

Several controls are multipurpose, which minimizes the total number of separate camera controls. The Power/Mode button performs both activities that the name suggests: it powers the camera on or off and selects a mode menu. The Shutter/Select button either starts or stops a video; takes individual photos or photo bursts; and starts or stops time-lapse photos. It also serves as a menu item selector in conjunction with the Power/Mode button, which selects the overall menu to be used.

Figure 9.3 is a diagram of the back of the Hero 3 camera that shows several features, including the important WiFi button, which only turns the WiFi on or off. All configuration of the wireless channel is done through the menu system. There is also a blue LED on the front of the camera that will flash if the WiFi link is on.

A Hero port that is available for plugging in certain accessories is also shown in Figure 9.3. A popular accessory is the LCD backpack, which is shown in Figure 9.4. The backpack makes it possible for users to view their photos and videos quickly without using a computer.

The Hero 3 has a simple optical system that is made up of a lens with a very small focal length, as shown in the Figure 9.5 diagram. This very small focal-length lens results in a very wide *field of view* (FOV), which is specified in the Hero 3 documentation as 170°. A lens with such an ultra-wide FOV is sometimes referred to as a *fisheye* lens, although the Hero 3 lens does not have the same extreme distortion as a true fisheye lens. Figure 9.6 shows several photos taken with a Hero 3 Black Edition in which this distortion is quite evident by the curved horizon. Technically, the fisheye distortion is known as *barrel distortion*.

The two original sample photos have also been post processed by imaging software that eliminates the inherent distortion. These photos are marked as "Corrected" in the figure.

FIGURE 9.1 GoPro Hero 3 camera.

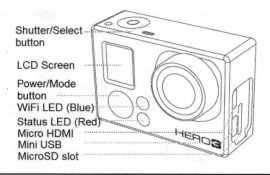

Shutter/Select button

LCD Screen

Power/Mode button
WiFi LED (Blue)
Status LED (Red)
Micro HDMI
Mini USB
MicroSD slot

Figure 9.2   Diagram of a Hero 3 camera front.

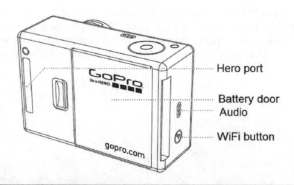

Hero port

Battery door
Audio

WiFi button

Figure 9.3   Diagram of a Hero 3 camera back.

Figure 9.4   LCD backpack.

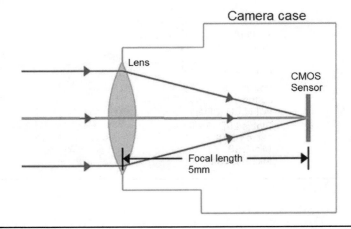

FIGURE 9.5   Hero 3 optical system.

Some people enjoy the mild fisheye distortion, since it adds a unique character to the Hero 3 photos. It should also be noted that it is probably impractical to correct any video distortion because of the enormous post-processing requirements that this would entail. Processing each video frame that is output at a 30 *frames per second* (FPS) rate would quickly result in a huge task.

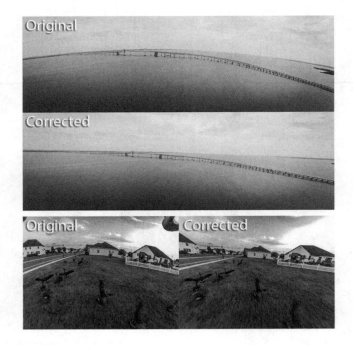

FIGURE 9.6   Hero 3 sample photos.

There is also the issue of choosing FOV versus close-up details, often referred to as telephoto. In Figures 9.7 to 9.11, you will see the series of photos I made in my back meadow. It illustrates the tradeoff between using FOV or telephoto with a Canon DSLR camera fitted with an 18- to 270-mm telephoto lens. The focal length is the only variable that was changed in each of the photos to illustrate how the FOV rapidly reduces as the focal length increases.

You can clearly see from this series of photos that the detail rapidly diminishes as the FOV expands. Figure 9.12 is another interesting example of this effect. It is an evening cityscape of Vancouver, BC, courtesy of Snapshot.com and photographer Darren Stone.

The Hero 3 optical system does not have any physical means to alter or change the focal length; hence, the optical FOV is fixed. It can, however, alter the FOV electronically by selecting different areas of the imaging sensor and expanding the selected pixels to fill the total image. This feature is shown in Figure 9.13, which shows the same scene taken on a Greek island in wide, medium, and narrow FOVs.

I next decided to test the Hero 3 camera to see how it handles very close-up photography. I used the classic Indian-Head test pattern that was widely used in the early days of monochrome, or black and white, television of the 1950s. This pattern is shown in Figure 9.14.

I took the photo shown in Figure 9.15 with the front of the Hero 3 lens only 4 cm (1.6 in) from the test pattern. You can clearly see the severe barrel distortion that happens when a Hero image capture is taken very close to the camera. The distortion lessens considerably as you separate the camera from the object being photographed. I also used the Adobe Photoshop lens-correction tool to see if the photo distortion could be somewhat mitigated. It was, to a small extent, as you can see in Figure 9.16.

Figure 9.7   Focal length of 18 mm.

Figure 9.8   Focal length of 50 mm.

Figure 9.9   Focal length of 100 mm.

FIGURE 9.10   Focal length of 200 mm.

FIGURE 9.11   Focal length of 270 mm.

FIGURE 9.12   Vancouver cityscape with various FOVs. *(Photo by Darren Stone, courtesy of Snapshot.com)*

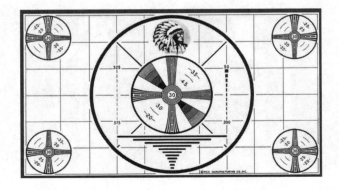

FIGURE 9.13   Greek island photos with different FOVs. *(Adapted from www.youtube.com/ watch?v=RUJ54EXjNCM)*

FIGURE 9.14   Test pattern image.

FIGURE 9.15   Hero 3 test-pattern image capture.

My only conclusion from the test result is that the Hero 3 is best used with a fair amount of distance between the lens and any object to be videoed or photographed. Fortunately, this will be the standard situation when the Hero 3 is mounted on the Elev-8.

Hero 3 WiFi-Range Test

I considered it to be very important to determine the maximum range from which the Hero 3 could be remotely operated with an Android tablet that is using the WiFi channel. WiFi range is dependent primarily on these three factors:

1. Environment
2. Protocol
3. Transmitter power

FIGURE 9.16   Photoshop post-processed Hero 3 test-pattern photo.

The environment is either indoors or outdoors. I selected the outdoor environment, since that is the area where I will usually operate the Elev-8. In addition, I will try to always have the quadcopter in my line of sight, since that also maximizes the range.

The second factor is related to the specific WiFi protocol that is to be used. The common protocols are IEEE 802.11a/b/g/n. The Hero 3 WiFi uses the Artheros AR6233 chip that supports all the common a/b/g/n protocols. The Android tablet I chose for the test has a "b" WiFi, which means that the maximum range is no more than 140 m (153.11 yd).

The last factor deals with the effective power radiated from the GoPro WiFi transmitter. This is impossible to know unless you dissect a GoPro, which is exactly what somebody did. Go to the website http://www.ifixit.com/Teardown/GoPro+Hero3+Teardown/12457/1 and see how the folks at iFixIt tore apart a Hero 3 Black Edition.

The actual test was very simple. I set the GoPro on a tripod in my back meadow and established a WiFi connection with an Android tablet that was running the free GoPro app. I clicked on *real-time preview* and saw myself on the tablet screen. I then moved away from the camera until I lost contact. This happened at about 100 m (109.36 yd) from the camera and tripod setup. This test was direct line of sight. I next walked back toward the setup to reestablish the communications link, which occurred at 100 m (109.36 yd). I took a picture at this point, which is shown in Figure 9.17.

I circled myself in the photo because it is very difficult to discern any details at that distance. I then walked 50 m (54.68 yd) toward the setup and took another picture, which is shown in Figure 9.18. This figure confirms both reliable communications at the maximum distance that I will be flying the Elev-8 and the maximum detail I could expect to see on the ground. Notice that I had to circle myself in the figure again, since it is still difficult to pick out distinct objects.

Figure 9.17   The 100-m (109.36-yd) range check.

FIGURE 9.18   50-m (54.68-yd) range check.

Ground Station

Ground station and *ground control station* (GCS) are general terms used to describe the means by which a quadcopter is controlled by a user. Normally, the GCS is just the R/C transmitter that an operator with visual contact uses to control the quadcopter. This approach becomes a bit more complicated when FPV capability is added. Now, the operator needs to have a video monitor available as an option rather than always maintaining a direct line of sight with the aircraft. Add in some telemetry, and the GCS can rapidly expand to become quite complex. My first attempt at creating a GCS will be to add a WiFi-enabled Android tablet to the Spektrum DX-8 transmitter and house both of them in an aluminum case. This setup allows the operator to use the DX-8 and simultaneously view the tablet screen while walking around. This simple GCS is shown in Figure 9.19 with a lanyard attached. This allows it to be easily managed in a hands-free manner.

The Spektrum telemetry is also displayed in the DX-8 LCD screen, which helps minimize the number of displays needed for this GCS. Using the DX-8 and Android tablet should suffice for most quadcopter operations that are held in a relatively restricted area. Operations in a wider area may become more problematic because it is very easy to lose situational awareness if you are out of line of sight with the quadcopter.

Incorporating the remote GPS data system described in Chapter 8 would help improve the operator's situation awareness. That system continuously transmits latitude, longitude, speed, and altitude back to a ground receiver. The received data is than displayed on an LCD

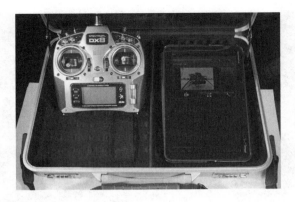

FIGURE 9.19   Simple GCS.

screen. The receiver, LCD display, BOE controller, and battery easily fit into the aluminum case shown in Figure 9.19.

A moving-map display from a laptop running Google Earth can show the quadcopter's position in real time. Relying on this display would be the ultimate way to improve operational awareness. However, this mode of operation would likely involve another operator who is available to continually input the latitude and longitude data that was being sent from the quadcopter. It would be too much of a workload for a single operator to both control the quadcopter and enter data, analogous to texting while driving.

Economy Video System

The FPV system creates excellent, high-definition video, but it is quite expensive. This section looks at a much less costly alternative that can provide adequate real-time video but does not have video-recording capability; however, I will discuss how such a capability can be provided for at the receiving end. This economy system is typically one-fifth to one-sixth of the cost of the GoPro system. Figure 9.20 shows the camera-transmitter module used in a

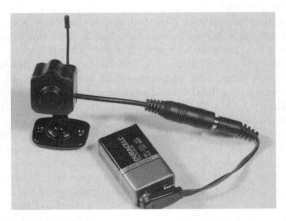

FIGURE 9.20   The RC310 camera-transmitter module.

FIGURE 9.21   Close-up of the RC310 camera.

system named the RC310. It is an inexpensive wireless 2.4-GHz camera system that is currently available from a number of online suppliers.

It is powered by a regular 9-V battery, as shown in the figure. It has an adjustable lens with a medium focal length, which is very different from the short focal length of the GoPro camera. Figure 9.21 is a close-up of the camera's front in which you can see the very small 2-mm diameter lens. The GoPro lens by comparison is 14 mm in diameter.

The small diameter of the RC310 camera lens means that the RC310 has a limited light-gathering capability. It will not perform well in poor lighting conditions; but this should not matter since the camera will normally be used in daylight conditions.

I took the photo shown in Figure 9.22 with the front of the RC310 camera lens 44.2 cm (17.4 in) from the Indian-Head test pattern. There is very little barrel distortion evident in the figure; however, the detail is quite fuzzy, which is due to the limited amount of pixels present in the camera sensor. Table 9.2 details the camera specifications.

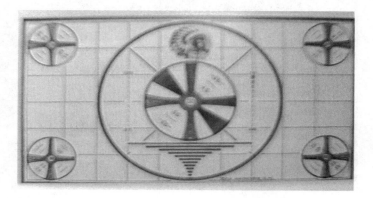

FIGURE 9.22   RC310 test-pattern image capture.

Features/Specifications	Description
Sensor	¼-in CMOS
Pixels	628 (H) × 582 (V) 510 (H) × 492 (V)
Dimensions	2.2 × 3 × 2 cm
Power requirement	5 V DC
Operating temp range	−20°C to +60°C
Signal system	NTSC
Scanning frequency	50 Hz or 60 Hz
Minimum illumination	3.0 Lux (F1.2)
Transmission range	150–200 m

TABLE 9.2 RC310 Camera Features and Specifications

The overall quality is essentially sub-VGA at 480 × 240 pixels, which means that this camera will not provide a detailed image when video is taken at a reasonable distance. But that is fine because this system was never designed for use with this type of application. I will show you later how to get some unusual and useful results from the video by using some clever post processing. Before getting to that, I want to discuss the receiver portion of the economy video system.

Figure 9.23 shows the receiver along with the video-capture module that is needed to capture the video frames for post processing.

Table 9.3 details the features and specifications for the RC310 receiver. The receiver was set to channel 1, which seemed to function well with little interference from other 2.4-GHz

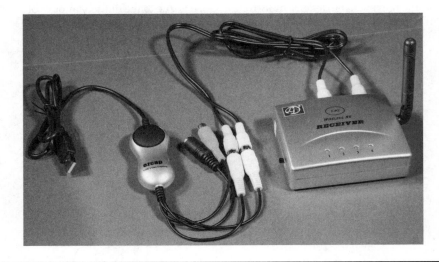

FIGURE 9.23 RC310 receiver and an EzCap USB video-capture module.

Features/Specifications	Description
Frequency	channel 1—2.414 GHz
	channel 2—2.432 GHz
	channel 3—2.450 GHz
	channel 4—2.468 GHz
Valid pixels	480 (W) × 240 (H)
Dimensions	9.6 × 7.9 × 3 cm
Power requirement	12 V DC
Operating temp range	0°C to +40°C
Modulation	GFSK
Audio output	10 kΩ/200 m Vp-p

TABLE 9.3   RC310 Receiver Features and Specifications

devices in the area. In addition, no configuration is required between the camera and the receiver, which is a very nice feature. A slide switch on the side of the receiver selects the operating frequency. The camera must also be set to the same frequency selected on the receiver. This is accomplished by setting two very small slide switches located on the back of the camera. These switches are shown in Figure 9.24.

The EzCap video-capture module, shown in Figure 9.23, is required for the interface between the RC310 receiver analog outputs and the laptop's digital USB input. Installing the USB driver is all that is required to use this video-capture device. It is totally transparent in its operation and shows up on the Windows Device Manager as a USB 2861 device in the sound, video, and game controllers category. This designation will be needed when you configure the post-processing software. Figure 9.25 shows how this versatile device can be used in a variety of different applications if you desire to do so.

FIGURE 9.24   RC310 camera frequency selection switches.

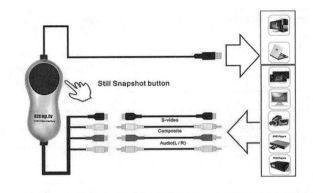

FIGURE 9.25   The EzCap video-capture configurations.

Post-Processing Software

The term *post processing* refers to subjecting a video or photograph that is already stored in a compatible file format to some type of image processing to enhance or extract desired features. This definition is best explained through an example, but first I will discuss the software used in post processing.

RoboRealm

A post-processing application that can be used in image analysis, computer vision, and robotic vision systems is named RoboRealm. Its startup screen is shown in Figure 9.26.

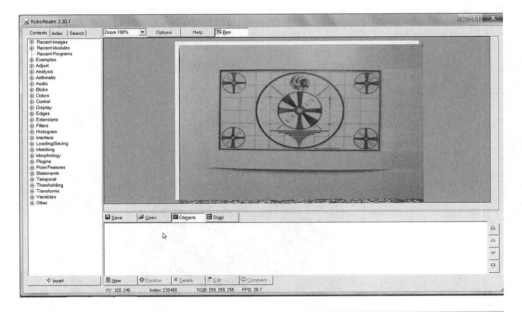

FIGURE 9.26   RoboRealm start-up screenshot.

Three main sections of this application deal with different aspects of image processing. The center section is where the processed image appears. The image shown in Figure 9.26 has not been processed, so it appears in its original, or raw, state. The section to the left is where you can select one or more processing options to apply to the image displayed in the center section. The bottom, center section shows, in sequence, all the processing effects that have been applied to the image. If it is applicable, parameter data will also appear in this section.

The RoboRealm application must be configured for the appropriate video source before any image processing takes place. Ensure that the EzCap device is plugged into a USB port before you start RoboRealm. Next, start the application, and click on the Camera button that is in the menu bar at the bottom of the center section. Then click on the Options button on the menu bar at the top of the center section, and select the Video tab when the Dialog box appears. A Camera Source selection textbox should appear as shown in Figure 9.27.

In my case, there were three selections shown. I chose the WDM 2861 Capture, which is the logical name for the EzCap device. That should do it, and you should see live video in the center section. At this point, you can actually process the live video stream if you desire. To demonstrate this, I chose a filter in the Canny edge detection operation from the left hand column and simply double clicked it. The results are shown in Figure 9.28.

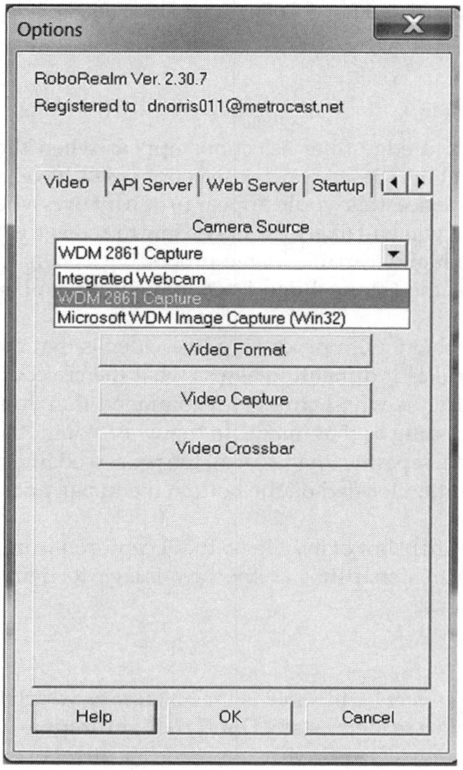

FIGURE 9.27   Camera Source selection textbox.

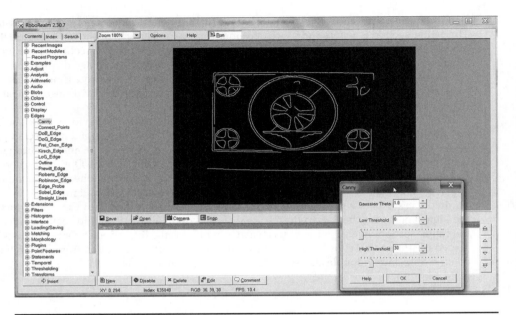

FIGURE 9.28   Live video filtered by a Canny edge algorithm.

Fourteen different edge filter selections appear when the + symbol is clicked on the Edges content selection. Other content selections will have multiple selections associated with them. At first glance, this would appear to be a bit overwhelming with so many different selections; however, you will likely use only a few to achieve what you want. The RoboRealm designers created a highly capable image-processing system that meets many different user needs. Incidentally, the Canny Edge Detector will be explained in greater detail later in this section.

An inherent problem with processing live video is that the original image is constantly changing, which makes it difficult to assess what the effect of multiple filter operations on an image will be. That is why I strongly recommend that you take a snapshot of the video and apply the processing to that image. In that way, you not only retain the original image, but you can also save separate images of all the processed images. To take a snapshot, simply click on the Snap button located on the bottom menu bar, and enter a name for the snapshot when prompted.

Figure 9.29 is a snapshot of my Elev-8 that I captured from a video taken with the RC310 camera. I will use this snapshot as the raw image to demonstrate a few of the image-processing algorithms.

The Histogram

The first processing effect I will show is the histogram, which is probably the algorithm that will be familiar to most readers. Many DSLRs have histogram displays to help photographers determine if their photos are properly exposed. The display is a small graph of vertical lines that show the distribution of all the *red, green,* and *blue* (RGB) pixel intensities contained

FIGURE 9.29   Elev-8 snapshot.

within the image. What you normally should look for is a distribution that is reasonably centered on the horizontal intensity scale, which in this case goes from 0 to 255. This range reflects the range values for eight-bit pixel intensity. Figure 9.30 shows the RGB histogram for the Elev-8 snapshot.

Outline Edge Extraction

The next effect is a little more complicated. I want to extract the Elev-8 outline from the raw image. This processing algorithm examines all the pixel values on a specific line in the image

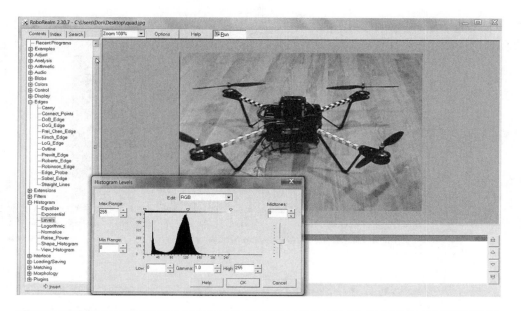

FIGURE 9.30   Elev-8 RGB histogram.

in a window and tests them to determine if the average intensity differs from a preset value or not. A radical change in value usually indicates that an edge is present at the center point where the windowed pixels are located. This algorithm is applied over all the lines that make up the image being processed. Figure 9.31 shows the Elev-8 snapshot processed to extract an outline.

You can also see in the figure that I selected a window size of 4 pixels, which is a good value. Any larger values will tend to make the edges less distinct, and smaller values will make the edges too intermittent and choppy. Selecting the correct parameters is almost always a process of trial and error, but that is in part why this field is so interesting.

Negative and Other Adjustments

Another interesting effect is the negative algorithm, an example of which is the negative snapshot of the Elev-8 shown in Figure 9.32. This effect is found in the Adjust Content category. The selections in the Adjust Content category are as follows:

- Camera_Properties
- Contrast
- Gamma
- Intensity
- Negative

Clicking on each one opens an additional variety of adjustments and settings that can be made to optimize a specific video-capture device. I was very impressed with RoboRealm's functionality with regard to camera setting. While I am not suggesting that RoboRealm

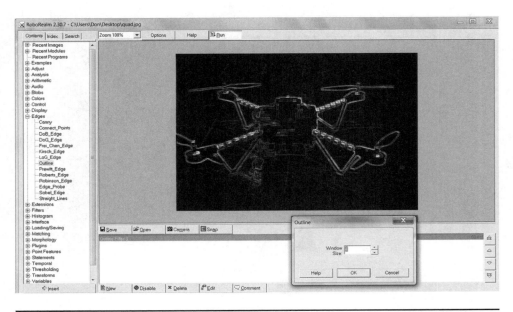

FIGURE 9.31   Elev-8 outline edge extraction.

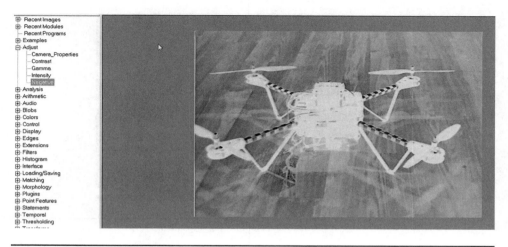

Figure 9.32   Elev-8 negative.

could replace Adobe Photoshop, I will say that RoboRealm seems to contain many of the same camera adjustments that are found in the more expensive Photoshop application as well as a few that are missing in the much more expensive program.

Canny Edge Detection

Edge detection is important because it facilitates shape detection and object separation. I selected the Canny edge detector as a representative image-processing algorithm that RoboRealm efficiently executes. This algorithm is named after John Canny who first published it in 1986. He wanted to create what he termed an optimal edge detector, one that would perform better than the many in existence at that time. His algorithm is a multistep one that I summarize below:

1. *Noise reduction*—This first step deliberately introduces a slight amount of blurring in the raw image through a Gaussian distribution that is convolved with the original image. The purpose is to eliminate or drastically reduce the effects of single noisy pixels.

2. *Determination of intensity gradients*—This next step determines the direction through which the edge or intensity gradient travels.

3. *Non-maximum suppression*—This step is essentially an edge-thinning technique. It uses an iterative approach that applies a 3 × 3 pixel filter to the edge that was determined in the previous step, which further refines that edge.

4. *Edge tracing and hysteresis thresholding*—This final step follows the intensity gradient, or edge, by ensuring that it is continuous or really the end of the edge. Once finished, every image pixel is marked as being either an edge or a non-edge.

The next two figures are from Wikipedia showing how the Canny edge detection works. Figure 9.33 is the original, unprocessed image of valves that are part of a steam engine. This figure contains many distinct edges and is a real test of the Canny algorithm.

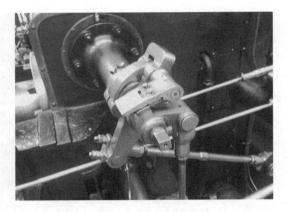

FIGURE 9.33   Steam-engine valves, the original image. *(Source: Wikipedia)*

Figure 9.34 shows the post-processed image for the steam-engine valves. I did notice that edge detection was not particularly successful in the portions of the image that had poor contrast. For instance, look at Figure 9.35 where I circled the vertical pipe coming out of the engine.

If you look at the same area in Figure 9.34, you can easily see the edges where the pipe exits the engine cover and turns downward. The curved edges are readily detected by the Canny edge detector. However, as the pipe descends, the contrast in the original image becomes much poorer, and the edges disappear, as is evident in Figure 9.34. It is just something to be aware of when post processing images. It is important to get as much light on the subject as possible and to maximize all edge contrasts for best results.

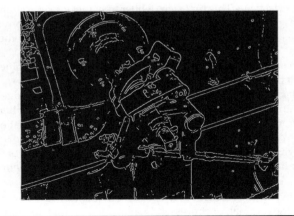

FIGURE 9.34   Canny edge detection on the steam-engine valves. *(Source: Wikipedia)*

FIGURE 9.35 Vertical pipe.

I also processed the original unprocessed image in Figure 9.33 through the RoboRealm Canny edge detector just to compare the results to the Wikipedia image in Figure 9.34. Figure 9.36 shows the result, which is very similar to Figure 9.34.

I will now show you the Elev-8 Canny-edge-detector image now that you know a little bit about this technique. Figure 9.37 is the processed Elev-8 image.

When I compared Figure 9.31, which is the outline technique, to Figure 9.37, it was apparent that the Canny approach produced more distinct edges than the outline approach did. The outline image seems to contain a substantial number of non-edge pixels, which I presume is a result of the simplistic, edge-detection algorithm that I discussed previously.

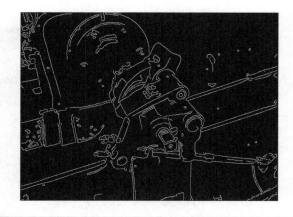

FIGURE 9.36 RoboRealm Canny edge detection on the steam-engine valves.

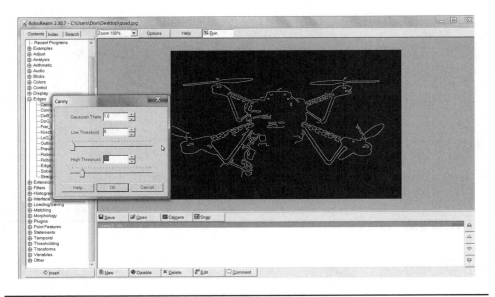

FIGURE 9.37   RoboRealm Canny edge detection on Elev-8.

Field Test of the RC310 System with Post Processing

To conduct a field test, I placed a bicycle near some vegetation and took some video to be post processed. Figure 9.38 shows the bicycle resting in some vegetation about 30 m (32.8 yd) away from the camera. The bicycle is identifiable in the image more by the blue color of the frame than by its shape. It still takes a bit of concentrated viewing to distinguish the bicycle from the background. Now, look at Figure 9.39, which is the same image but after it has been post processed by applying the outline-edge-detection algorithm.

FIGURE 9.38   Test image of the bicycle.

FIGURE 9.39   Test image of bicycle after outline edge detection.

Now you should be able to see the fence in the foreground, the trees in the background, and the bicycle frame. I was fairly sure that the bicycle wheels would not be detectable because of the poor contrast and limited pixel resolution. It is important to recognize that the frame shape is out of context with its surroundings, thus making it easier to detect. By out of context, I am referring to the vertical trees and fence sticks as well as the horizontal fence parts. The frame is conspicuous because it has edges that are neither vertical nor horizontal and are contiguous, or close together. This edge characteristic makes the object identification a bit easier and is one of the cornerstones that experts in airborne video surveillance constantly use.

I also decided to apply the Canny edge detection to the test image of the bicycle. Figure 9.40 is the result. Believe it or not, although the bicycle frame is in the image, it is just about invisible because of the default Canny parameters that were initially used. After some parameter adjustments, I was able to obtain the results shown in Figure 9.41.

Yes, the only edges showing in the figure belong to the bicycle frame! It really is quite amazing what can be accomplished when you use clever image-processing techniques. Admittedly, I had to *play around* with the parameters until I achieved this remarkable result. The Gaussian Theta setting had to be changed from 1.0 to 1.5. This setting occurs in step 1 in the Canny algorithm that deals with pre-blurring. The other change I made was to raise the High Threshold from 30 to 83. The threshold settings occur in step 3 in the algorithm.

Higher Resolution Test Image

I decided to repeat the test-image experiment with a DSLR image instead of one from the economy camera. My goal was to determine what effects image quality would have on edge recognition. Unsurprisingly, it turned out that image quality has a significant effect on edge detection. Figure 9.42 is another picture of the bike, this time taken with a Canon 40D equipped with a 70- to 200-mm telephoto lens.

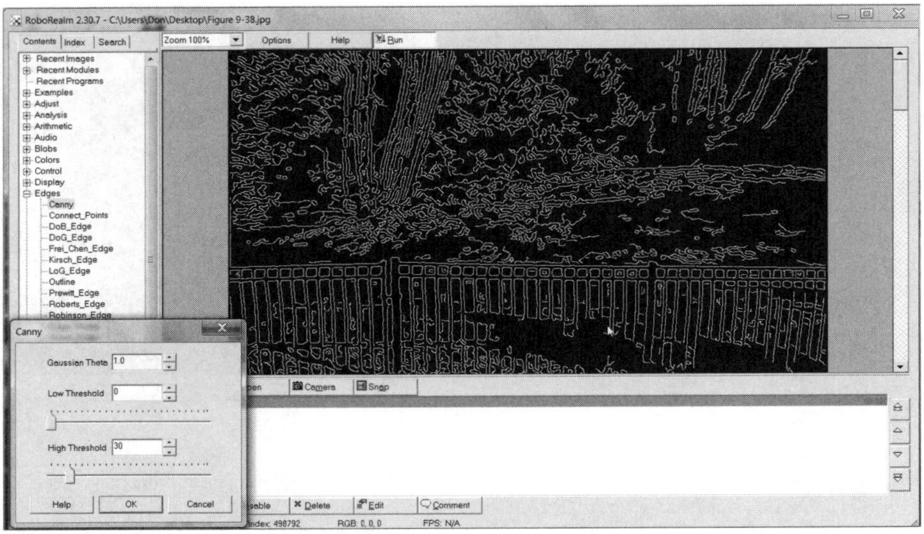

Figure 9.40   Test image of bicycle after Canny edge detection.

This image has been significantly resized to match the economy image. I applied the outline edge detector first, as I had done in the first test. Figure 9.43 is the result of this outline-edge-detector post processing. This time you can easily see the tire outlines and even the outlines of the handlebar and seat—a definite improvement over the previous outline edge test.

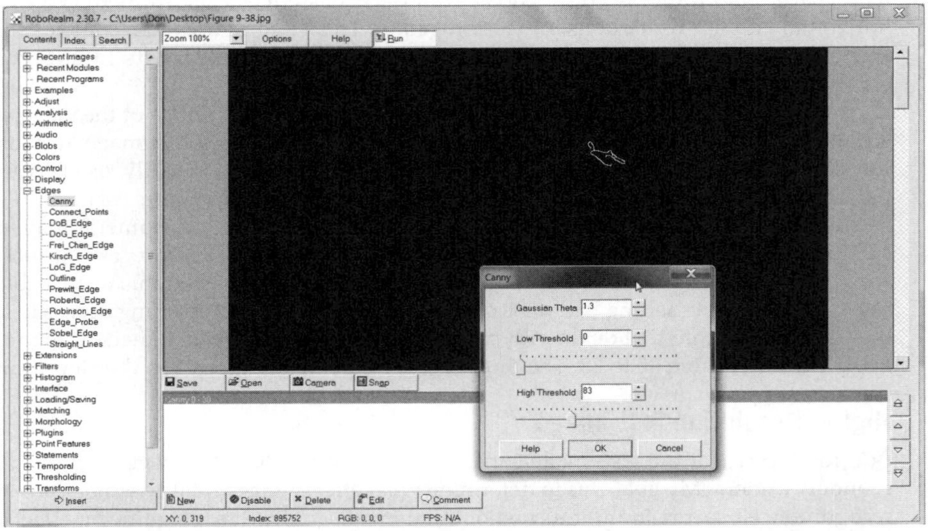

Figure 9.41   Test image of bicycle after adjustment of Canny edge detection.

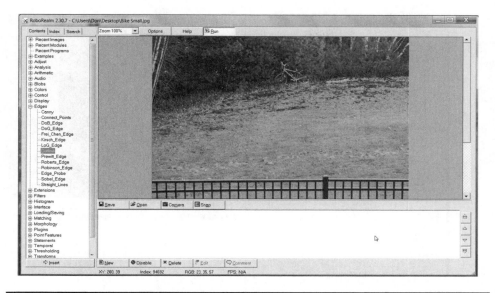

Figure 9.42   High-resolution test image.

The Canny edge detection was next. I had to experiment with the parameters to extract the best edges. The result may be seen in Figure 9.44. This time you can observe the two tires, the handlebar, and if you squint a bit, the seat. There are more non-target object edges in this image, which I believe is due to a greater number of pixels that the algorithm attributed to pseudo-edges.

Figure 9.43   Outline edge detection of the high-resolution test image.

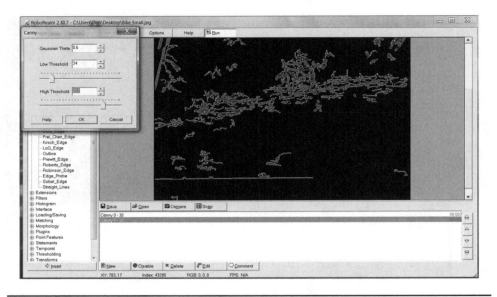

Figure 9.44   High-resolution test image adjusted by Canny edge detection.

Based upon the results of the above tests, I believe that the ideal camera for use in surveillance and targeted-object detection would be a GoPro format type of camera with a longer optical focal length. I am not aware of the availability of any such camera to consumers, but it would be a good choice for this type of surveillance. I am sure that the tradeoff between lens size and camera-body size is a big consideration for designers and marketing people for this video camera type.

Geotagging GoPro Hero 3 Photos

Geotagging refers to inserting GPS coordinates into a photograph's metadata. Metadata in turn refers to the hidden data included in a digital image file that provides a great deal of supplemental information concerning the photographic image. *Exchangeable image file format* (Exif) is the name for this image metadata and is a photographic industry standard that has been in existence since the mid 1990s. Most modern digital cameras automatically generate Exif data, which is appended to the actual image data. Exif data is usually viewed through applications designed to reveal the data, although it is easily viewable in the Windows operating system by following these three simple steps:

1. Right click on the image name in the directory where it is stored.
2. Click on Properties, which is normally the bottom selection in the Dialog box.
3. Click on the Details tab.

You should be able to scroll through the Dialog box and view all the metadata that has been stored with regard to the selected image. You may even have geotagged photos on your computer if they were taken with a smartphone with location service enabled.

CAUTION: Photos that were taken with a smartphone and posted on the Web can inadvertently provide GPS coordinates to anyone with an interest. You should probably avoid this if you and your family are leaving on a road trip or vacation and take a parting photo at your home.

Geotagging is appropriate only for photographs, not video, since it is impossible to tag every frame when they are being created 30 times per second. The Hero 3 camera has a useful feature that takes a still photograph image every 5, 10, 30, or 60 seconds, while also creating a video. These intermittent photos will be the ones geotagged to record your location on the photo. Notice, that I chose the word "your" to describe the location because the GPS coordinates will be generated by the Android tablet that is part of the ground control station (GCS) and not by anything onboard the quadcopter. In most cases, this should not be a problem, since you will be within 100 m (109.36 yd) of the flying quadcopter. The GCS GPS coordinates should be sufficiently accurate for most location and identification purposes.

Geotagging photos require that a record of GPS coordinates be saved during the same time interval that the photographs are taken. This saved record is known as a *GPS track*, which is just a collection of GPS coordinates along with the time that the coordinates were taken. It then becomes a simple task of matching the time the photograph was taken to the matching time from the GPS track. Most digital photos have the time they were taken recorded and stored in the Exif data. I used a program named OpenGPS Tracker to create the GPS track that runs on a Motorola Android Zoom tablet. This tablet is part of the GCS shown in Figure 9.19 on page 244.

Geotag Test Run

I decided to run a simple experiment in which I made a three-mile trip near my home with the Hero 3 camera mounted in my car. Figure 9.45 shows the Hero 3 mounted on the front windshield with a suction-cup mounting adapter.

If you examine the figure closely, you will see that the camera is mounted in an inverted position. A camera setting in the Setup menu will invert the image to compensate for this common mounting configuration. Note that the Hero 3 is also mounted in an inverted position on the tiltable platform.

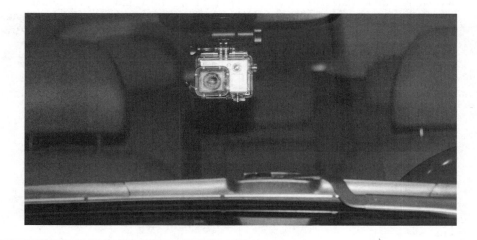

FIGURE 9.45   Hero 3 mounted on a car windshield.

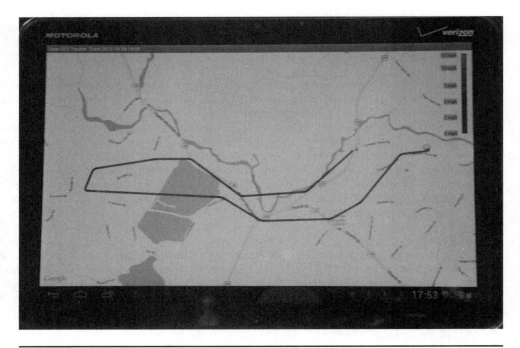

FIGURE 9.46   The test-run track on the Android screen.

At the start of the trip, I clicked on the Start tracking button in the Android tablet's OpenGPS Tracker application. Then I periodically took pictures throughout the trip and finally clicked on Stop Tracking at the trip's end. The tracker application created a track log, a file named Track201310231000.gpx, that will be one of the inputs to the geotagging program. Figure 9.46 shows the track as it appeared on the tablet's screen.

The gpx extension appended to the track file is short for *GPS Exchange Format*. This format is in the *extensible meta language* (XML), which allows for standardized data interchange between applications and Web services. The gpx file can hold coordinates, also known as waypoints, along with routes and tracks. The following is a snippet of the track log:

```
<?xml version='1.0' encoding='UTF-8' standalone='yes' ?>
<gpx version="1.1" creator="nl.sogeti.android.gpstracker"
xsi:schemaLocation="http://www.topografix.com/GPX/1/1
http://www.topografix.com/gpx/1/1/gpx.xsd"
xmlns="http://www.topografix.com/GPX/1/1"
xmlns:xsi="http://www.w3.org/2001/XMLSchema-instance"
xmlns:gpx10="http://www.topografix.com/GPX/1/0"
xmlns:ogt10="http://gpstracker.android.sogeti.nl/GPX/1/0">
<metadata>
<time>2013-10-26T14:00:25Z</time>
</metadata>
<trk>
<name>Track 2013-10-26 10:00</name>
```

```
<trkseg>
<trkpt lat="43.23794960975521" lon="-71.04583740234169">
<ele>61.0</ele>
<time>2013-10-26T14:00:26Z</time>
<extensions>
<ogt10:accuracy>9.487171173095703</ogt10:accuracy></extensions>
</trkpt>
<trkpt lat="43.237617015837365" lon="-71.04948520660194">
<ele>35.0</ele>
<time>2013-10-26T14:02:16Z</time>
<extensions>
<ogt10:accuracy>45.59917449951172</ogt10:accuracy></extensions>
</trkpt>
<trkpt lat="43.23302507400387" lon="-71.05446338653358">
<ele>-23.0</ele>
```

XML employs user-generated tags to delimit the data. An example from the above data snippet is:

```
<time>2013-10-26T14:00:26Z</time>
```

where `<time>` is the beginning tag and `</time>` is the ending tag. Everything between the tags is data that can be extracted by an XML parser application. XML is a versatile data interchange format that is rapidly becoming the mainstream way to tranfer data in most software applications and Web services.

The second input file needed for geotagging is the one containing all the images. I created this second file by removing the microSD card from the Hero 3 camera and mounting it on my Windows laptop. You will probably need an adapter to hold the microSD card so that it can be inserted into a standard laptop SD card port. Figure 9.47 shows a typical adapter that is often provided when you purchase a microSD card.

FIGURE 9.47 A microSD card adapter.

I then copied all the images created during the trip from the card into a directory that I named GoPro test. One of the images in this directory is shown in Figure 9.48. The figure shows some of the Exif data for this image. I retrieved the data by using the three steps described above. I am showing you this data so that you can see what it contains before it is geotagged.

The free Windows program I used to geotag the photos is named GPicSync. It can be downloaded from https://code.google.com/p/gpicsync/. Figure 9.50 shows the application's main screen.

You have to enter the following information before geotagging. Your specific directory, file, and time zone information will differ from mine.

- *Pictures folder*—C:\Users\Don\Pictures\GoPro Test\GoPro Test
- *GPS file*—C:\Users\Don\Downloads\Track201310261000.gpx
- *Google Earth* — > Icons—picture thumb   Elevation—Clamp to ground
- *Check*—Create a log file in pictures directory
- *Check*—Add geonames and geotagged—Geonames in IPTC + HTML Summary in IPTC caption
- *Select time zone*—US/Eastern
- *Geocode picture only if time difference to nearest track point is below (seconds)*—300

Click on the Synchronize! button to start the geotagging process. It might take a while to complete the process, especially if there are a lot of photos to be tagged.

FIGURE 9.48   Sample trip image.

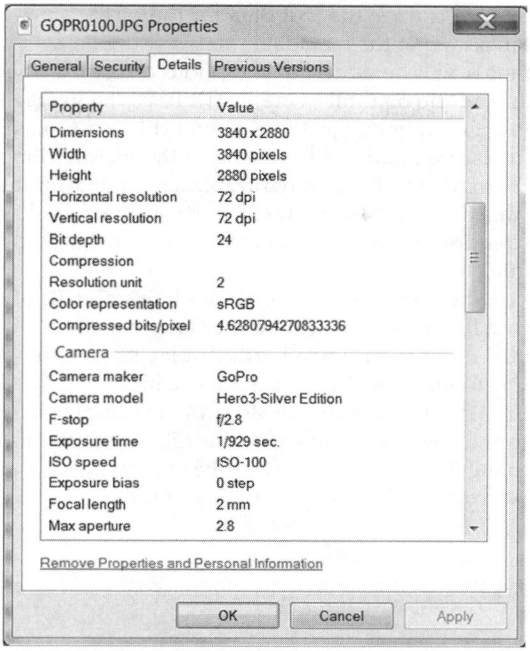

FIGURE 9.49   Exif data for the image in Figure 9.48.

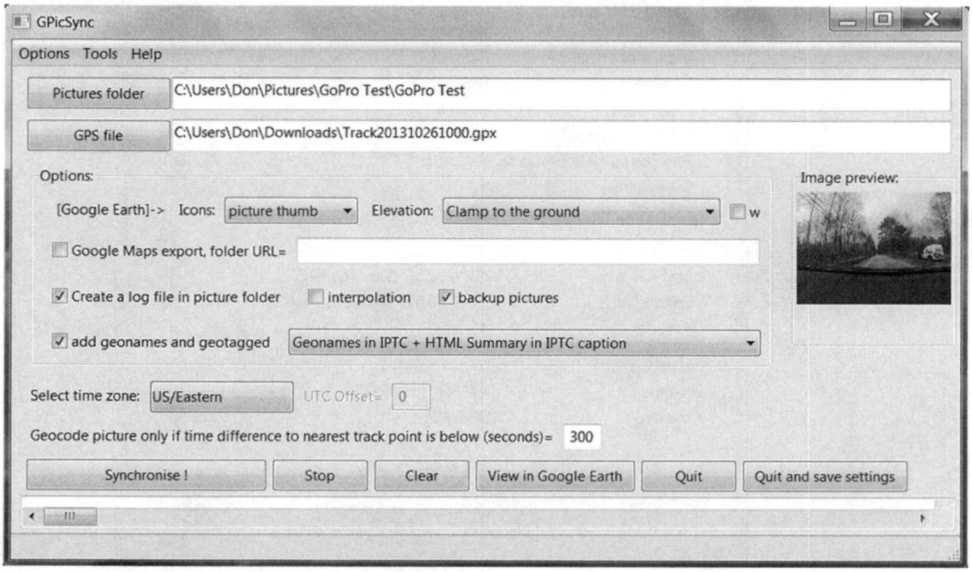

FIGURE 9.50   GPicSync main window.

In Figure 9.51, you can see the Exif data for the earlier sample photo now showing that GPS data has been added to the metadata.

Duplicate photos will be added to the photo directory. The first one will be the non-geotagged photo, and the one following it will have the geotag data added. You will now have two versions of each photo, just in case you still want to use an untagged photo.

There should also be another file created in the picture directory named doc.kml. This file is in the XML format, which is a recognizable data source for the Google Earth application. Double click on this file and choose Google Earth as the default application to open this file with the .kml extension. Figure 9.52 shows the result that will occur when Google Earth opens with this file.

As you can see, the track is shown along with thumbnails of the photos that were taken when I was driving the track. I clicked on the sample picture to show where it was taken in the track. I zoomed in on the Google Earth display so that the photo could be more easily seen; however, the zoom limited the track view a bit.

A snippet of the doc.kml file that deals with the sample photo is shown on the opposite page. It is easy to see how the sample photo is identified. In this case, it is GOPRO100.JPG, which is between the XML name tags. The GPS coordinates are also included in the snippet and are shown between the XML coordinates tags.

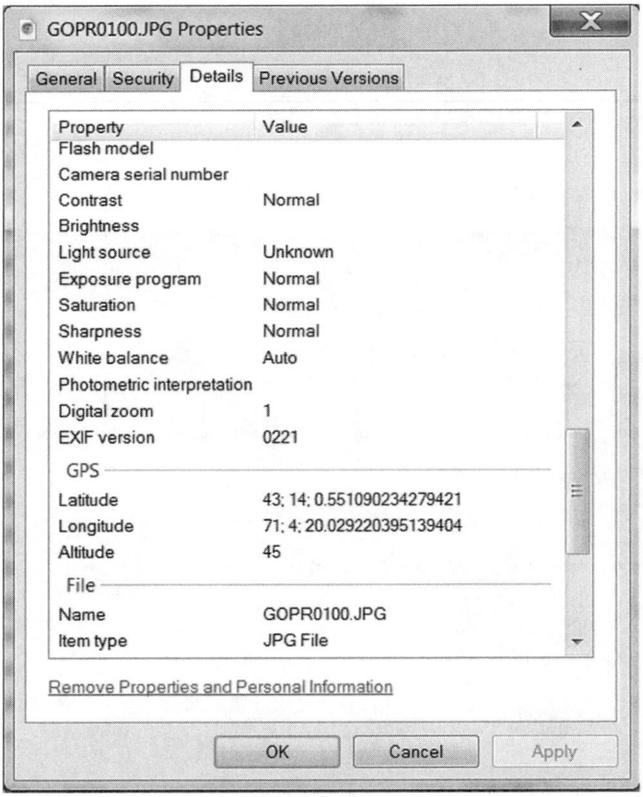

FIGURE 9.51   Geotag data added to Exif data for image in Figure 9.48.

FIGURE 9.52   Google Earth opened with the doc.kml file.

```
<Placemark>
<name>GOPR0100.JPG</name>
<description><![CDATA[<img src='gopr0100.jpg' width='600'
height='450'/>]]>
</description>
<styleUrl>#defaultStyle1</styleUrl>
<Style>
<IconStyle>
<Icon><href>thumbs/thumb_GOPR0100.JPG</href></Icon>
</IconStyle>
</Style>
<Point>
<coordinates>-71.0458740234169,43.23794960975521,61.0
</coordinates>
</Point>
</Placemark>
```

An easier way to view the track and photos than the above procedure is to simply click on the View button in Google Earth, which is located on the GPicSync screen. I initially discussed the doc.kml file because I wanted to show you what was really happening when you clicked this button.

This section concludes my discussion on the video systems that I have found effective for quadcopter operations. The next chapter focuses on performance checks, along with some important training topics.

Summary

The chapter began with an introduction to the GoPro Hero 3 camera system. This was one of two video systems discussed in the chapter. The Hero 3 is a very wide angle, high-definition video system that is well suited for implementing a *first-person viewer* (FPV) system that helps the operator to control the quadcopter.

I discussed the Hero 3's basic functions along with its limitations so that you have a clear understanding of its capabilities. A WiFi range check was also conducted to confirm that the Hero 3 should operate quite well with the quadcopter.

I next discussed a simple *ground control station* (GCS), which allows the operator to have all the communication devices readily available for flying the aircraft in a safe and effective manner.

Next, we looked at an economy video system that provides a reasonable surveillance capability when used in conjunction with post-processing software. I demonstrated the RoboRealm software application, which processed video captures from the economy video system. I showed you how to perform histogram, negative, and edge-detection algorithms with the economy video system. I also provided a detailed explanation of Canny edge detection and its algorithm to illustrate how you could go into great depth with one of the many RoboRealm algorithms.

I showed you the results of two field tests, one using the economy video system and the other conducted with a DSLR.

The chapter concluded with a discussion of how to geotag a series of Hero 3 photographs. I also showed you how to create the GPS track file that is required before you can geotag the photos.

CHAPTER 10

Training Tutorial and Performance Checks

Introduction

In the first part of this chapter, I will discuss how you should train to safely operate a quadcopter. I will also recommend various resources that are available to enhance this training. The last part of the chapter concerns quadcopter performance measurements. I discuss the various factors that can either increase or diminish a quadcopter's operating performance.

Developing Fundamental Quadcopter Piloting Skills

First, I am not suggesting that the only way to fly a quadcopter successfully is to use a training program to develop some pilot skills. However, you have probably invested considerable time, effort, and money in building your quadcopter and probably should want to protect your investment. Developing good operating skills takes a modest investment in both time and money but is considerably better than the alternative trial and error system, especially where error might cost you your whole quadcopter investment or worse, such as injury to others and/or property damage. With this disclaimer expressed, I will proceed to discuss the actual training that I accomplished.

Training to operate the quadcopter is best accomplished by practicing with a software simulator program. There are a variety of quadcopter simulator programs that I have found and that are listed in Table 10.1. This list is not guaranteed to be totally comprehensive or accurate as of the time you read it. The R/C software industry is as dynamic as most software development companies, and new programs appear and old ones disappear quite frequently.

In my discussion from this point on, I will be using the AeroSIM RC flight simulator, which I will now simply refer to as the "SIM." The SIM is available online from www .aerosimrc.com and is moderately priced. It comes with a training cable that is discussed below. It supports the following connector outline styles that are shown in Figure 10.1 and detailed in Table 10.2. The SIM has many features including but not limited to:

- Realistic scenarios
- Game mode
- Sixty-three lessons, including takeoff, flight controls, and landing

Name	Remarks
Real Flight RF7	Comprehensive with two quadcopter models included; moderate cost.
Real Flight Basic	A less expensive version of the above simulator; however, it does not come with a quadcopter model. A free user-created model is available from Knife Edge Software.
RC Phoenix Pro V4	Seems to be highly regarded, judging from user reviews.
Flying Model Simulator (FMS)	This program was developed by Draganfly Innovations. No quadcopter models are included, but several free ones are available from the Draganfly Forum.
AeroSIM RC	The one that I use and will discuss in this chapter.

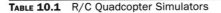

TABLE 10.1   R/C Quadcopter Simulators

- Reset to takeoff position after a crash
- Realistic sun glare (not too applicable to quadcopter operation)
- Realistic graphics
- GPS included in *first-person video* (FPV)
- Ability to create your own scenarios using satellite imagery
- Model editor
- Instrument panel (not too applicable to quadcopter operation)
- Comprehensive menu bar at the bottom of the main screen
- Wind adjustments

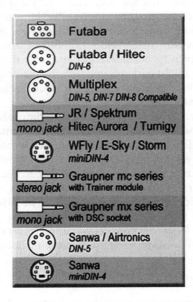

FIGURE 10.1   SIM connector outlines.

Name	Manufacturer
Futaba	Futaba
DIN-6	Futaba/Hitec
Multiplex	No specific maker—DIN-5, DIN-7, DIN-8 compatible
Mono jack	JR/Spektrum/Hitec Aurora/Turnigy/Graupner mx series with DSC socket
Stereo jack	Graupner mc series with Trainer module
Mini-DIN-4	WFly/E-Sky/Storm/Sanwa
DIN-5	Sanwa/Airtronics

TABLE 10.2   Supported SIM Connectors

- Power, maneuverability, and responsiveness settings
- Record and replay flights
- Transmitter-to-USB cable included

The SIM is extremely comprehensive and includes all the features that are needed to make you a competent quadcopter operator. It is moderately priced and, I believe, well worth the cost. Figure 10.2 shows you what comes with the SIM.

One of the most important aspects of the SIM is that you control the quadcopter with your actual R/C transmitter and not with the mouse and keyboard. This mode of operation brings an important dose of realism to the training that is directly transferable to actual

FIGURE 10.2   SIM package contents.

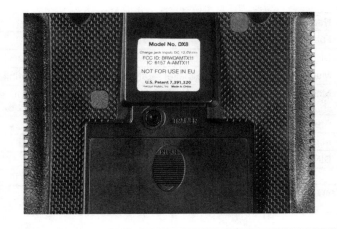

FIGURE 10.3   Spektrum DX-8 Trainer jack.

quadcopter operations. I think it is worth further looking into to discover how the actual R/C transmitter signals are incorporated into the SIM. Figure 10.3 shows the 3.5-mm trainer jack that is located on the back of the Spektrum DX-8 R/C transmitter.

> **NOTE:** *Some R/C transmitters label this jack as the buddy box because it provides the necessary signals that allow another transmitter to work alongside the primary, or main, transmitter. I discuss buddy-box training later in this chapter.*

All eight *pulse-position modulation* (PPM) signals are available when a 3.5-mm plug is inserted into the trainer jack. These PPM signals represent the eight transmitter channels and are generated every 22 ms, as can be seen in Figure 10.4.

The timing for these signals was thoroughly discussed in Chapter 6, and I would urge you to go back and review that discussion to fully comprehend what these pulses represent. There is one significant difference between the pulse trains shown in Chapter 6 and those that are emitted from the trainer jack. In the ones under discussion here, the actual pulses themselves are fixed in width at 0.4 ms each. It is the time between the pulses that represents the actual channel-pulse width. It really is just how you interpret the trace.

You should note that the DX8 transmitter should be off when the 3.5-mm plug is inserted. There is a switched contact built into the jack that will automatically turn the DX8 on and place it into the Slave mode when a plug is inserted. Also, note that the transmitter is disabled while it is operating in the Slave mode.

I also expanded the time scale on the USB oscilloscope because I wanted to confirm that the throttle channel was properly functioning. Figure 10.5 shows the expanded trace with all eight channel pulses shown and with the throttle at the 0 percent setting.

The throttle pulse-width spacing is 1.1 ms, which is the expected value. I next increased the throttle to approximately 80 percent in order to check the new pulse width. All other channels remained unchanged. Figure 10.6 shows the result.

The new throttle pulse-width spacing is approximately 1.7 ms, which matches the 80 percent setting. Consequently, I assumed that all the other channels were functioning properly.

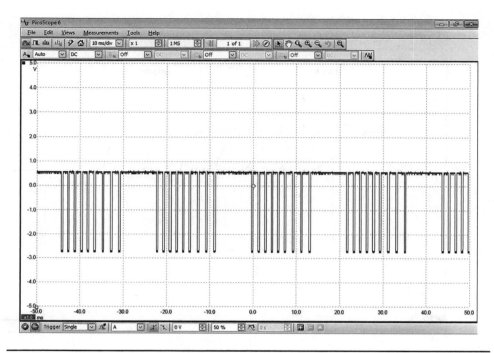

FIGURE 10.4 USB oscilloscope trace from the trainer jack output.

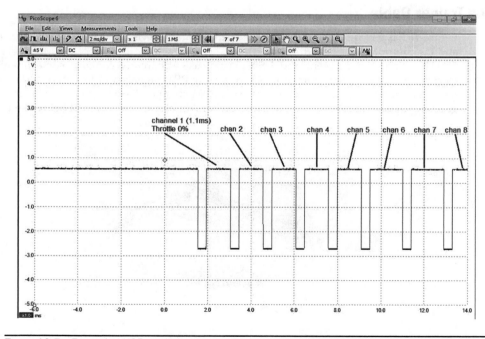

FIGURE 10.5 Expanded USB oscilloscope trace with throttle at 0 percent.

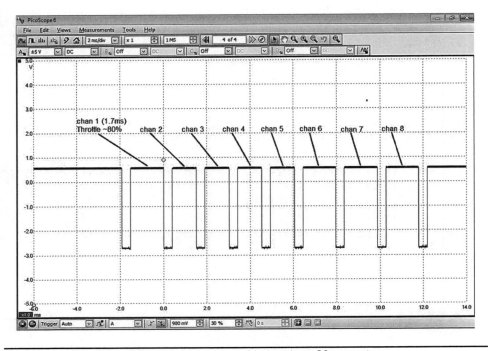

FIGURE 10.6   Expanded USB oscilloscope trace with throttle at 80 percent.

The Trainer Cable

Figure 10.7 is a close-up photo of the SIM cable that connects between the DX8 trainer jack and the computer's USB port.

The slightly translucent, red USB plastic case apparently contains some interesting electronics that are programmed to accept analog PPM signals and to output serial digital data that represents the data value for each PPM channel. Based upon my research, I am guessing that some type of low-cost, low-power Atmel microprocessor is being used in the

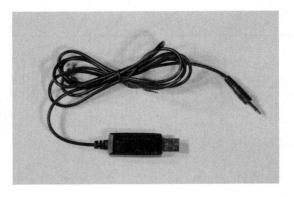

FIGURE 10.7   SIM trainer cable.

adapter case that determines the timing between the fixed pulses and outputs a packet of eight numbers representing that time. This is repeated every 22 ms to match the PPM stream from the transmitter. These data packets are subsequently received and processed by the SIM to provide the user with control inputs to the application.

It would be relatively easy to program a computer to supply similar data packets, and thus, give you a way to experiment with various control inputs without needing an R/C transmitter. In fact, there are a number of folks who have done exactly what I just mentioned so they could experiment with SIMs and other devices that accept PPM inputs.

The SIM installation CD contains the required device driver that must be installed before you can use the trainer cable. Note that the SIM is only Windows-compatible as of the time of this writing.

Running the SIM

It is not possible in this book to really show you how the SIM functions, but presenting some screenshots of it working will help you understand. I will show a few figures that are representative of what you would see when running the SIM. Figure 10.8 is the opening screen that you see when the SIM starts. I had already selected the quadcopter IV model, which is shown in the figure. You must configure the SIM to match the transmitter's controls. The easiest way to accomplish this important step is to follow the procedure below:

1. Click on the Controller selection located in the top menu bar
2. Click on the Configuration selection
3. Click on the Controls button
4. Click on either Config A or Config B (I selected Config B)
5. Click on the Config and Calibrate TX (Beta) button
6. Follow the step-by-step instructions

FIGURE 10.8 The SIM opening screenshot.

FIGURE 10.9   SIM screenshot with quadcopter flying.

The SIM should now be all set up to operate after completing the above steps. The model is flown in the SIM, using the DX8 controls in exactly the same way they would be operated if you were using the actual quadcopter. There is also a simulated FPV view that mimics, to some extent, the real GoPro FPV. Figure 10.9 shows a quadcopter FPV with GPS, system status, wind, and course data superimposed on the SIM screen. This display is great for SIM operation but unfortunately is not present for real operations.

Figure 10.10 shows a typical lesson in which you have to keep the quadcopter hovering within the circle. In this lesson, you will learn that very small control movements are all that you need to maneuver the quadcopter. You will quickly lose control if you make large or

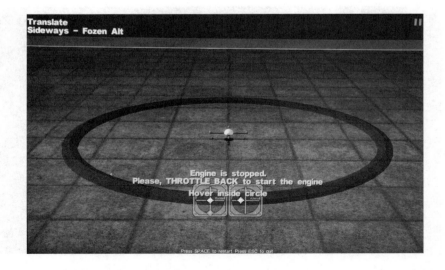

FIGURE 10.10   Sample SIM lesson screenshot.

even moderate control-stick motions. It is important to realize that finesse and a slight control motion is all that is needed.

The SIM keeps track of your progress throughout all the lessons and is a handy way to rate your own progress. I do not believe you have to complete all the lessons to simply become reasonably proficient in operating the simulated quadcopter. I would recommend that you complete the fundamental lessons, such as maintaining altitude and hovering within a designated location, even though you may already have some R/C experience. Controlling the quadcopter SIM is significantly different from a conventional airplane.

The Buddy Box

The phrase *buddy box* refers to the situation in which two R/C transmitters are wired together to control a single aircraft. Figure 10.11 shows a buddy-box cable that is used to connect two Spektrum R/C transmitters.

The plug ends of the cable are inserted into each transmitter's trainer jack to complete the buddy box. One transmitter must be configured as a Master and the other as a Slave. The Master unit is the only one that transmits to the aircraft, while the Slave unit passes its control inputs to the Master, which then transmits them. The Slave operator can control an aircraft while being observed by the Master operator. The Master operator can immediately take control of the aircraft if the Slave operator inadvertently places the aircraft in a bad or dangerous flight condition.

The configuration instructions appropriate for each transmitter need to be followed in order to get the buddy box working. It is not required that each transmitter be the same model or even from the same manufacturer. It will probably require some trial-and-error experiments to get different manufacturers transmitters working in a buddy-box arrangement.

The following instructions show you how to set up a buddy box between two DX8 transmitters:

1. Ensure that both transmitters are using the same aircraft profile. You can accomplish this by using an SD card to copy the setup from one DX8 and then loading that setup into the other DX8. Be sure that the same model is selected in both transmitters.
2. Press and hold the roller while powering up the DX8 that you designate as the Master. Rotate the roller and select the Trainer mode. Then select Pilot Master from the available options. Finally, exit and power off.

FIGURE 10.11   Buddy-box cable.

3. Press and hold the roller on the Slave or Student DX8 while powering up the unit. Rotate the roller and select the Trainer mode. Then select Slave from the available options. Finally, exit and power off.

4. Now, turn on the Master DX8 and plug in the trainer cable. Do not turn on the Slave DX8. Simply plug the trainer cable into the Slave DX8, which will power it on. The term SLAVE should appear in the Slave LCD screen.

5. Remove all the propellers from the quadcopter before proceeding with this step. Next, test that the Master allows the Slave to power the quadcopter and that the motors appear to change speed as the controls are moved.

6. Ensure that no motor-speed changes occur when control is shifted between the Master and Slave and vice-versa.

Wireless SimStick Pro

This section discusses an enhancement to the basic SIM setup. I used a device named SimStick Pro that enables a totally wireless interface between the R/C transmitter and the laptop running the SIM. This SIM enhancement also requires the use of an R/C receiver along with the PPM-to-USB cable previously discussed. Figure 10.12 shows the setup with the SimStick Pro, PPM-to-USB cable, and a Spektrum AR8000 satellite receiver.

All the components shown in the figure are powered from the laptop's USB port. I did find that I had to rebind the satellite receiver to the transmitter in order to establish a communications link. Refer back to Chapter 6 to refresh your memory regarding the binding process.

This setup not only breaks the cable link between the transmitter and the laptop, but it also uses an actual radio link to further promote a realistic operating configuration. Using this configuration has absolutely no effect on how the SIM functions. It just allows you to walk around with the transmitter exactly as you would in the field. It would also be a distinct

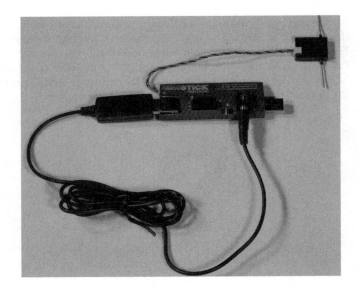

FIGURE 10.12 Wireless SimStick Pro setup.

advantage if the laptop video could be connected to a large-screen, flat-panel monitor or television so that you would not have to be close to the laptop screen. That would comprise an ideal training situation.

Performance Measurements

Measuring quadcopter performance can mean different things, depending on your viewpoint and needs. I think that measuring payload, or lifting capability, beyond the quadcopter's own weight would be an important item on everyone's list. After all, such capability is a fundamental property for any quadcopter performing a task such as video surveillance. Another key property is flight time, or stated another way, determining how long a quadcopter can stay aloft before the battery charge depletes and it must be landed. An additional property that might be of interest to a select few users is the maximum altitude at which the quadcopter can fly above ground. For most users, this is irrelevant because local regulations likely restrict the maximum altitude to 400 ft (121.9 m) or less.

Determining Maximum Payload

I decided to focus on payload as the property that would interest most quadcopter users. Maximum payload, as I am defining it, is the total weight of a fully loaded quadcopter less the weight of an unloaded one but with a battery attached. Determining maximum payload capacity has probably been done in a trial-and-error fashion in which a payload is attached to the quadcopter and a flight trial is attempted. The quadcopter would either fly or not, and if it did fly, it might tip upside down if it happened to be at its maximum weight capacity. That could lead to serious quadcopter damage that might even include damaging an expensive video camera if it happened to be the payload. I first tried tethering or tying down a quadcopter using tough string, but that never worked, as the quadcopter flight-control system would "fight" the tether and would always twist to one side or another. I figured there must be a safer and more scientific way to determine payload. Figure 10.13 is a sketch of my first design for determining payload.

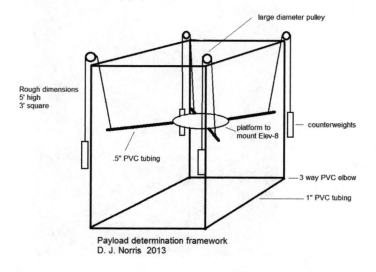

FIGURE 10.13   Preliminary design sketch for maximum payload determination.

The basic concept was to mount the quadcopter on a platform that is counterbalanced to match the combined weight of the platform and an unloaded Elev-8. In that way, I could add additional weight to the platform until the Elev-8 could no longer lift the weight. That weight must then equal the maximum payload.

Figure 10.14 is a photo of the completed test frame. Several revisions were made to the original design before I arrived at the one shown in the photo. I have included the plans for this test frame on this book's website, www.mhprofessional.com/quadcopter, for those readers interested in building the device. I deliberately used common materials that are typically found in home improvement stores to make it easy to build it.

The top of the test frame is shown in Figure 10.15. You can see the common wash-line pulleys I used to redirect the nylon line that holds the platform to the hanging counterweights.

The platform, which is supported by a cross brace made of 0.5-in PVC tubing, is shown in Figure 10.16. In this figure, you can see that the empty platform is being held tight to the travel-stop collars by the hanging counterweights. These stop collars are required to prevent the Elev-8 propellers from striking the top-frame cross member as the quadcopter rises to the top. You should also notice the PVC tees that hold the ends of the platform cross brace to the vertical frame tubes. These tees were significantly altered to allow free vertical travel, while still providing minimal horizontal motion. Figure 10.17 is a close-up of one of these PVC tees.

Figure 10.14   Final test-frame design.

Figure 10.15   Top view of the test frame.

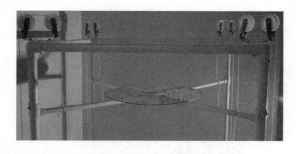

Figure 10.16   Close-up of the test frame platform.

Figure 10.17   Platform cross-brace attachment tee.

Each tee was bored out using a 1⅜-in Forstner bit to allow enough clearance for it to slip over the 1½-in OD vertical PVC tube. A 90° arc was also cut out of the main barrel to prevent binding, yet allow the tee to be pushed onto the vertical tube.

The hanging counterweights are pictured in Figure 10.18. These counterweights are made up of 2½-in OD PVC tubes with a hard cap on one end and a rubber cap on the other end. Simple friction holds the hard cap in place, while a 3-in hose clamp secures the rubber cap. This arrangement allows water to be put into each counterweight until the appropriate weight is reached. The following calculations show how the water weight was determined for the counterweights.

Total weight of platform and Elev-8:

$$1.140 \text{ kg (Platform wt.)} + 1.664 \text{ kg (Elev-8 wt.)} = 2.804 \text{ kg}$$

Counterweight calculations:

Average weight of one empty counterweight = 0.560 kg

Total weight of four empty counterweights:

$$0.560 \text{ kg} \times 4 = 2.240 \text{ kg}$$

Total weight of water needed to counterbalance weight of the platform and the Elev-8:

$$2.804 \text{ kg (wt. of platform \& Elev-8)} - 2.240 \text{ (wt. of empty counterwts.)} = 0.564 \text{ kg}$$

Weight of water to be added to each counterweight:

$$0.564 \text{ kg} \div 4 = 0.141 \text{ kg}$$

Weight of each counterweight with water added:

$$0.560 \text{ kg} + 0.141 \text{ kg} = 0.701 \text{ kg}$$

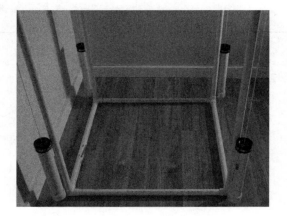

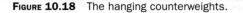

Figure 10.18   The hanging counterweights.

FIGURE 10.19   The Elev-8 mounted on the test frame.

Figure 10.19 shows the Elev-8 mounted on the frame platform for payload testing. Notice that the Elev-8 is positioned so that the propellers are centered between the vertical tube supports. This positioning ensures that the propellers have ample clearance from any frame member.

I attached the Elev-8 to the platform using tie wraps, as shown in Figure 10.20. In order to use the tie wraps you will need to drill two holes on either side of the landing-gear touchdown points, as you can see in the figure.

Test Results

I ran a series of experiments by adding a little bit of weight each time until the quadcopter began to vibrate excessively, thus indicating that it had reached its lifting capability. The result was approximately 0.9 kg (1.98 lb). I estimated that this value should vary by ± 50 g,

FIGURE 10.20   Elev-8 tie-wrap attachment to the platform.

depending upon the distribution of the load. This means that the Elev-8, as configured, should have a conservatively rated 0.9-kg (1.98 lb) payload, which should meet most users' needs.

I also noted that the Hoverfly flight-controller LED would start flashing red when the quadcopter started to shake violently, which happens at the maximum capacity. I observed that one of the motors began to overheat to the point where I could smell burning insulation. Fortunately, the motor recovered when allowed to cool; however, I now suspect that the motor will not perform at optimal capacity because of the overheating event. (Refer to my motor discussion in Chapter 5 to see the reasons for my pessimism.) This kind of behavior should not be present in unconstrained free flight, since the flight controller is functioning as designed and will not direct massive power to a single motor in order to correct perceived force imbalances. I believe that the quadcopter should function normally when it is in free flight because the flight controller will receive appropriate responses to its control commands to the motors.

Kill Switch

At the end of Chapter 3, I mentioned the desirability of adding a kill switch to the Elev-8. This feature would simply cut off all primary power to the quadcopter and cause it to drop out of the sky before it could crash into people or property. Figure 10.21 shows a simple kill-switch configuration. It is just a toggle switch that can be turned off by actuating a servo outfitted with a simple wooden arm that trips the switch.

NOTE: Use an appropriately rated toggle switch that can handle the peak current that flows from the primary battery during typical quadcopter operations.

I connected the servo directly to the uncommitted Aux-2 channel on the AR-8000 R/C receiver. Flipping the Aux-2 switch from the + (positive) position to the − (negative) position

FIGURE 10.21   Elev-8 kill switch.

causes the servo arm to rotate 90°, which is sufficient to turn the kill switch on. The quadcopter will instantly lose all power and immediately drop when the kill switch is activated. Of course, be very sure that no one or no thing is under the quadcopter as it drops. The whole point is to prevent injury or property damage, not to cause it by carelessly dropping the quadcopter.

Estimating Flight Time

Flight time is related directly to the LiPo battery's *state of charge* (SOC). The only parameter that is measured while in flight is the battery voltage, which is reported back to the DX8 transmitter via the telemetry module. The main problem is how to relate battery voltage to SOC. A solution to this problem is not too simple because the battery voltage and SOC are not directly proportional. In other words, this means a 20-percent reduction in battery voltage does not mean that the SOC is 20-percent depleted because the relationship between voltage and SOC is a complex, or nonlinear, one.

R/C enthusiasts have done studies relating battery voltage to SOC, which I summarize in Table 10.3. I show both single-cell (1S) and three-cell (3S) voltages in this table.

On average there is a 10 percent SOC depletion for every 5-mV LiPo-cell voltage drop. I would caution you that these values may not represent the exact behavior of your LiPo battery, as there are significant variations in the way LiPo batteries are manufactured for the R/C field. However, the data should be reasonably close and more than suitable to address the concerns of most quadcopter operators.

There is a general rule of thumb that applies to LiPo batteries that you must know. It is the so-called 80-percent rule:

CAUTION: Do not drain the battery below 80% of its rated charge.

An 80-percent depletion, according to the data in Table 10.3, reflects a 3S voltage of 11.4 V. This means you should immediately land the quadcopter if the telemetry shows you are approaching 11.4 V. Although there are complex reasons why LiPo batteries should not

Single Cell LiPo Voltage	3-Cell LiPo Voltage	Percentage SOC
4.20	12.60	100
4.15	12.45	90
4.10	12.30	80
4.05	12.15	70
4.00	12.00	60
3.95	11.85	50
3.90	11.70	40
3.85	11.55	30
3.80	11.40	20
3.75	11.25	10
3.70	11.10	0

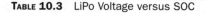

TABLE **10.3**   LiPo Voltage versus SOC

be discharged below 80 percent of rated capacity, I will simply state that following the 80 percent rule will allow you to achieve the maximum battery life possible. Most consumer LiPo's can be recharged 350 to 400 times before losing their recharge capability. Discharging past 80 percent of capacity will significantly reduce the number of recharge cycles. This can be an expensive situation, since high-capacity LiPo batteries are not cheap.

I would like to relate battery voltage to flight time now that I have thoroughly discussed LiPo SOC. The key parameter is current flow, or discharge rate. The motors probably account for 90 to 95 percent of the current flow. The amount of current the motors draw is directly related to the payload and flight dynamics. The current flow can be minimal if the quadcopter is idling on the ground with the propellers either not moving or just barely rotating. The flow is significantly higher while it is hovering and likely peaking if you are racing across the countryside or seeking a new high-altitude record.

I will use an average current flow of 16 A for each of the A2212/13T motors to reflect the operational modes that a quadcopter operator will likely use in a typical flight. I will also use a 40C, 3S battery as the power source, which again might be a very typical choice. The total average current draw for all four motors would be 64 A, which is quite a large energy drain. However, modern LiPo batteries are quite capable of meeting this demand, which is the primary reason why quadcopters can operate.

The flight time calculation is:

$$40C/64 \text{ A} \times 60 \text{ min} = 37.5 \text{ min } (100\% \text{ depletion})$$
$$37.5 \text{ min} \times 0.80 \text{ depletion} = 30 \text{ min } (80\% \text{ depletion per the } 80\% \text{ rule})$$

Thirty minutes would be the absolute maximum that you could expect; however, in practical terms, I believe you should expect realistic flight times in the range of 22 to 28 minutes. Of course, if you use a lower capacity battery, your flight times will be proportionally less. You should expect no more than 15 minutes of flight time if you use a 20C battery.

My practical advice to quadcopter operators is to have several fully charged LiPo batteries available, in addition to the one mounted on the quadcopter. In that way, you can quickly swap batteries and not have to waste time waiting for the depleted battery to recharge, which can be a considerable interval, especially if it is down to its minimum SOC.

This completes the discussion on training and quadcopter performance, I will now move on to another interesting topic: where the quadcopter field is heading (no pun intended). Chapter 11 will also include suggested ways to improve or experiment with your quadcopter.

Summary

I started this chapter with a discussion of how to acquire the skills needed to successfully operate a quadcopter. Operating an R/C simulator program seems to be the most prudent way to develop the required skills, while minimizing any potential damage to the real quadcopter. Several available simulator programs were listed, with particular focus on the AeroSIM RC (SIM) program, as that was the one I had purchased to train myself.

The discussion that followed looked at the various connector styles that are commonly used to hook up the R/C transmitter to the computer running the SIM. A 3.5-mm mono plug is used with the Spektrum DX8 transmitter that I chose for this project.

I showed you a series of USB oscilloscope traces that demonstrated the *pulse-position modulation* (PPM) signals that are available at the trainer jack located on the back of the DX8. I also discussed the PPM-to-USB converter module that changes the PPM signals to the equivalent serial digital numbers that are input into the computer's USB port. These numbers are used by the SIM to represent the user's control motions as sensed by the R/C transmitter.

Several screenshots were shown to illustrate typical SIM displays that you will see when you are running the actual SIM.

I briefly discussed the buddy-box training scenario in which two operators could control a real quadcopter. One operator would be designated as the Master and the other operator as the Slave. Presumably, the Slave operator would be the one learning to operate the quadcopter, while the Master operator could override and take control if something was going awry.

The wireless SimStick Pro accessory was discussed next. It allows the SIM to be operated in a very realistic manner whereby the transmitter sends the control signals to the R/C receiver by means of an actual radio link. From that point, it goes through the PPM-to-USB module/cable and into the SIM computer. This arrangement is about as close to flying a real quadcopter as you can practically achieve.

The chapter closed with sections that dealt with two quadcopter performance measures: payload and flight time. I showed you a unique test frame that I designed and built to accurately measure the actual payload a quadcopter could lift. *Payload* is normally defined as the amount of weight that can be lifted while not counting the actual quadcopter weight including the battery. In the case of the basic Elev-8, I determined the payload was 0.9 kg (1.98 lb). This is sufficient to lift most loads that a medium-sized quadcopter such as the Elev-8 could expect to lift, given that a fairly hefty LiPo battery was powering the aircraft.

The last section dealt with determining flight time. I went through a detailed discussion regarding LiPo *state of charge* (SOC) and battery voltage showing that it is possible to reasonably estimate the SOC simply by viewing the real-time battery voltage. That measurement is available if you have incorporated the TM1000 telemetry module in the onboard electronics. I also went over the 80 percent rule, which is critical to follow to ensure that the LiPo battery life can be extended by its maximum number of possible recharges.

CHAPTER **11**

Enhancements and Future Projects

Introduction

In this chapter, I will discuss both enhancements and future projects that you may want to consider for your quadcopter. Several of the concepts that were introduced in earlier chapters will now be integrated into these discussions. You might want to go back and refresh your knowledge regarding these concepts as you read about them. I have endeavored to point out the appropriate chapters in which they were first discussed. You will also find some new material about advanced sensors that will add significant flexibility and capability to your quadcopter flight operations.

Position Location and Return to Home Operation

Determining the quadcopter's geographic position is relatively easy using the GPS system. In Chapter 10, I described and demonstrated a simple, real-time GPS system that continuously sent the GPS coordinates back to the *ground-control station* (GCS). These coordinates were then displayed on an LCD screen in a latitude and longitude format. In that discussion, I showed how these coordinates could be manually entered into the Google Earth program to provide a real image of where the quadcopter was positioned in the nearby terrain. GPS data can also be used onboard the quadcopter to direct the aircraft to travel to a previous location or to a new location. Unfortunately, I cannot show you how to implement a GPS positioning system onboard the Elev-8, since the flight-control software in the Hoverfly Open flight-control board is proprietary (as I mentioned in Chapter 3). I can, however, outline a proposed approach on how to implement a virtual system that will position a generic quadcopter by using GPS data.

This generic quadcopter will be a very simple one consisting of four motors driven by four *electronic speed controllers* (ESCs) that are, in turn, controlled by the *virtual flight-control system* (VFCS), which is implemented by the Parallax *Board of Education* (BOE). For this project, I will assume that the VFCS will respond to R/C control signals from the GPS module and also from an electronic compass. Figure 11.1 shows a block diagram of my generic quadcopter control system that I just described.

In Chapter 10, I described the onboard GPS module. In the new configuration, the GPS module's signals will be connected directly to a BOE instead of a Prop Mini. In addition,

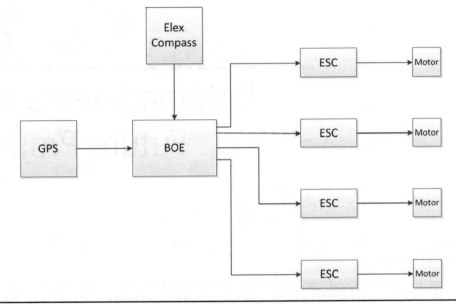

FIGURE 11.1 Block diagram of a generic quadcopter control system.

there will be another sensor input from an electronic compass that I describe and demonstrate in the following section.

Electronic Compass Module

I use the Parallax Compass Model HMC5883L with the Parallax part number 29133. This module is described as a sensitive, three-axis electronic compass with sensors that are very capable of detecting and analyzing the Earth's magnetic lines of force. This module is shown in Figure 11.2 and is quite small and compact.

FIGURE 11.2 Parallax Compass Module HMC5883L.

FIGURE 11.3   Honeywell three-axis compass sensor.

The module uses the basic physics principle of *magnetoresistance* in which a semiconductor material changes its electrical resistance in direct proportion to an external magnetic flux field that is applied to it. (The effect was first discovered in 1851 by William Thomson, who is more commonly known as Lord Kelvin.)

The HMC5883L utilizes the Honeywell Corporation's *Anisotropic Magnetoresistive* (AMR) technology that features precision in-axis sensitivity and linearity. These sensors are all solid state in construction and exhibit very low cross-axis sensitivity. They are designed to measure both the direction and the magnitude of Earth's magnetic fields, from 310 milligauss to 8 gauss (G). Earth's average field strength is approximately 0.6 G, well within the range of the AMR sensors. Figure 11.3 is a close up of the three-axis sensor used in the compass module.

Three magnetoresistive strips are mounted inside the module at right angles to one another in a way that enables them to sense the X-, Y- and Z-axes. It is probably easiest to picture the X-axis aligned with the Earth's north-south magnetic lines of force, which then makes the Y-axis an east-west alignment. The Z-axis can now be thought of as the altitude, or depression. Figure 11.4 shows these axes superimposed on the Earth's lines of force.

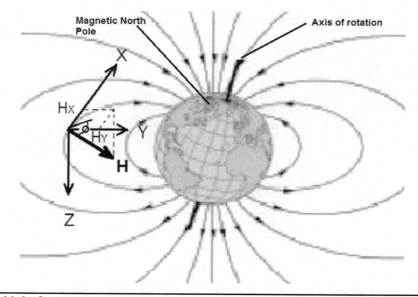

FIGURE 11.4   Compass axes superimposed on the Earth's magnetic lines of force.

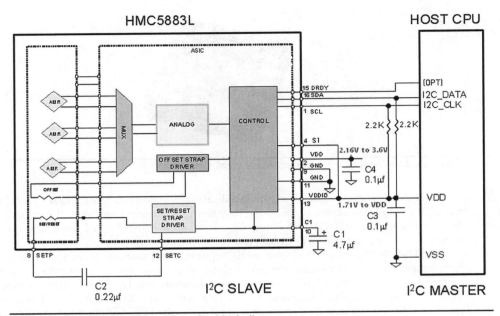

FIGURE 11.5 The HMC5883L electronics block diagram.

The figure is a bit hard to decipher, as there is a lot being shown. The bold line labeled **H** represents a magnetic line of force vector that can be decomposed into three smaller vectors that are aligned with the X-, Y-, and Z-axes. I have shown only the Hx and Hy components, for clarity's sake.

The angle between the plane formed by Hx and Hy components and the **H** vector is called the *declination* and is usually represented by the Φ symbol. Note that it is also important to compensate for the tilt of the compass in order to achieve an accurate bearing.

Fortunately, all the compensation and calculations are nicely handled for us within the HMC5883L module. Figure 11.5 shows a block diagram of what constitutes the main electronic components within the module.

The module communicates with the host microprocessor using the I²C bus that was discussed in Chapter 6. It takes a total of only four wires to connect the compass module to the BOE. These connections are detailed in Table 11.1. The physical connections can be seen in Figure 11.6.

We now need software to test the compass module with the BOE. I downloaded a very nice demo program named *HMC5883.spin* from the Parallax OBEX website, which I have

HMC5883L Pins	BOE Pins
VIN	5 V
GND	GND
SCL	P0
SDA	P1

TABLE 11.1 Connections between the HMC5883L Module and the BOE

Figure 11.6 Physical connections between the HMC5883L module and the BOE.

discussed in previous chapters. This program contains all the test code necessary to demonstrate the compass functions as well as the I²C driver software that communicates between the compass module and the BOE. In addition, the program uses the FullDuplexSerial object, which provides a terminal display using the Propeller Serial Terminal (PSerT). This program is also available from this book's website, www.mhprofessional.com/quadcopter.

The first portion of the test program with comments is shown below.

```
{{HMC5883.spin 2011 Parallax, Inc. V1.0
Controls a Honeywell HMC5883 3-axis compass over an I²C bus.
Demo shows raw X, Y, Z and calculated Azimuth plus a heading in
Degrees.


 ┌─────────────┐
 │ H    SDA    │    ───    I²C Data pin, I²C Master/Slave Data
 │             │                (Data I/O)
 │ M    SCL    │    ───    Serial Clock — I²C Master/Slave Clock
 │             │                (Clock 160 Hz)
 │ C    DRDY   │    ───    Data Ready, interrupt pin. Internally
 │             │                pulled high. (opt.)
 │ 5    VIN    │    ───    2.7 - 6.5 V DC (module is regulated to
 │             │                2.5 V DC)
 │ 8    GND    │    ───    Ground
 │ 8           │
 │ 3   -Module │
 └─────────────┘
}}
```

```
CON
  _clkmode    =  xtal1 + pll16x
  _clkfreq    =  80_000_000
  datapin     =  1                    'SDA
  clockPin    =  0                    'SCL


" All available registers on the HMC5883 are listed below:


  WRITE_DATA    = $3C    'Used to perform a Write operation.
  READ_DATA     = $3D    'Used to perform a Read operation.
  CNFG_A        = $00    'R/W Register, Sets Data Output
                          Rate. Default is 15Hz
                          8 samples per measurement. 160Hz can be
                          achieved by monitoring DRDY.
  CNFG_B        = $01    'R/W Register, Sets the Device Gain(230-
                          1370 Gauss). Default = 1090 Gauss.
  MODE          = $02    'R/W Register, Selects the operating mode.
                         'Default = Single measurement.
                         'Send $3C $02 $00 on power up to  change
                          to continuous measurement mode.
  OUTPUT_X_MSB  = $03    'Read Register, Output of X MSB 8-bit
                          value.
                         'Will read -4096 if math overflow during
                          bias measurement.
  OUTPUT_X_LSB  = $04    'Read Register, Output of X LSB 8-bit
                          value. Will read -4096 if math overflow
                          during bias measurement.
  OUTPUT_Z_MSB  = $05    'Read Register, Output of Z MSB 8-bit
                          value. Will read -4096 if math overflow
                          during bias measurement.
  OUTPUT_Z_LSB  = $06    'Read Register, Output of Z LSB 8-bit
                          value. Will read -4096 if math overflow
                          during bias measurement.
  OUTPUT_Y_MSB  = $07    'Read Register, Output of Y MSB 8-bit
                          value. Will read -4096 if math overflow
                          during bias measurement.
  OUTPUT_Y_LSB  = $08    'Read Register, Output of Y LSB 8-bit
                          value. Will read -4096 if math overflow
                          during bias measurement.
  STATUS        = $09    'Read Register, indicates device status.
  ID_A          = $0A    'Read Register, (ASCII value H)
  ID_B          = $0B    'Read Register, (ASCII value 4)
  ID_C          = $0C    'Read Register, (ASCII value 3)

VAR
long x
```

```
long y
long z

byte NE
byte SE
byte SW
byte NW

OBJ
    term        :       "FullDuplexSerial"      'PSerT driver
    math        :       "SL32_INTEngine_2"      'Math library needed for
                                                 the atan function

PUB Main
  waitcnt(clkfreq/100_000 + cnt)                'Wait while compass has
                                                 time to startup.

  term.start(31, 30, 0, 9600)                   'start a terminal Object
                                                 (rxPin, txPin, mode,
                                                 baud)

  setcont                                       'sets continuous data
                                                 acquistion

  repeat                                        'Repeat indefinitely
    setpointer(OUTPUT_X_MSB)                    'Start with Register
                                                 OUT_X_MSB

    GetRaw                                      'Gather raw data from
                                                 compass

    term.tx(1)                                  'Set Terminal data at top
                                                 of screen

    RawTerm                                     'Terminal window display
                                                 X,Y,Z Raw Data

    HeadingTerm                                 'Terminal window display
                                                 of heading in degrees.

PUB HeadingTerm
  "Terminal window display of heading in degrees.
    term.str(string("Heading in Degrees:",11))
    term.tx(13)
    term.tx(13)
    Heading

PUB AzimuthTerm
  "Terminal window display of calculated arcTan(y/x)
    term.str(string("This is the calculated azimuth:",11))
    term.tx(13)
    term.tx(13)
    term.str(@Azm)
```

FIGURE 11.7 Operational demonstration circuit.

```
term.dec(azimuth)
term.tx(13)
term.tx(13)
```

Much of this program code simply controls how the data is displayed on the PSerT. The actual program used in the VFCS will be much more condensed, since there is no need for a human readable display to be implemented.

Figure 11.7 shows the demonstration circuit in operation. I have included a traditional compass in the figure to illustrate magnetic north. The whole BOE has been oriented such that the compass module is pointing to magnetic north. Figure 11.8 is a PSerT screenshot confirming that the compass module is indeed pointing north.

It is time to examine the latitude and longitude calculations now that I have established how to measure real-time magnetic bearings. These background discussions will set the foundation for your understanding of how quadcopter operations tasks, such as return to home base, can be accomplished.

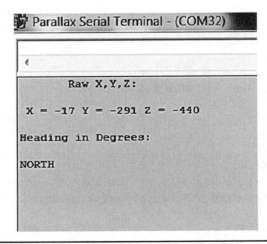

FIGURE 11.8 PSerT screenshot displaying the north direction.

Computing Path Length and Bearing Using Latitude and Longitude Coordinates

I need to review some fundamental principles regarding latitude and longitude that you might have missed if you snoozed during fifth-grade geography class. The horizontal lines in Figure 11.9 are lines of latitude and the vertical lines are lines of longitude. Lines of longitude are also known as meridians, and lines of latitude are sometimes referred to as parallels.

If you were to cut the Earth along a line of longitude, it would result in a circle, from which you would conclude that the planet is a spheroid, but it is not. However, for our purposes, I will make the reasonable assumption that the Earth is a sphere, since the computed distances in quadcopter operations are miniscule with regard to the radius of Earth. Any errors introduced by this assumption are too small to be realistically determined.

All the cross-sectional circles cut on longitude lines have the same diameter, which makes path determination easy for true north-south travel directions. Cutting along the horizontal latitude lines, however, results in circles that diminish in size, which you should be able to envision readily as you travel from the Equator (the maximum-diameter circle) to either the true North or South Poles (minimum-diameter circles). This change in diameter greatly complicates distance determination, but it is handled by a series of calculations (shown later in this section).

All the latitude and longitude circles are further divided into degrees, minutes, and seconds to establish the geographic coordinate system. The *zero degree* (0°) line of longitude is defined by international standards to run through Greenwich, England. Vertical lines to the left of the 0° line are designated as west while lines to the right are designated as east. The lines continue to the opposite side of the Earth ending at 180° for both east and west lines of longitude. The 0° line of longitude is also known as the *Prime Meridian*.

Figure 11.10 shows how the distance between lines of longitude narrows as it travels north from the 0° line of latitude, which is also known as the *Equator*. The same holds true for traveling south from the Equator. Table 11.2 shows precise measurements of longitudinal arcs at selected latitudes.

The ±0.0001° measurement is of the most interest for quadcopter operations because it will be the typical scale for flight operations. This means that the degree measurements of latitude and longitude must be accurate to at least four right-hand-side decimal points.

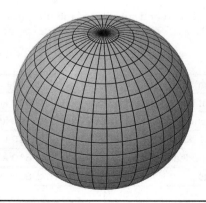

Figure 11.9 Lines of latitude and longitude.

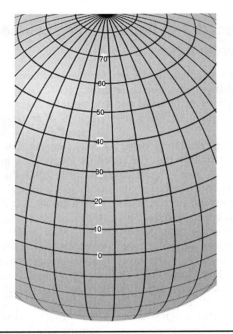

FIGURE 11.10   Lines of latitude and longitude.

The distances between lines of latitude, or parallels, will be the same no matter at which specific latitude they are measured. For instance, at the Equator, a 1° of latitudinal length is 111.116 km, which is almost exactly the same as 1° of longitudinal length. The following equation holds true for all latitudinal length calculations:

$$\text{Latitudinal length} = \pi \times M_R \times \cos(\varPhi)/180$$

where:
$M_R = 6367449$ m (Earth's radius)
$\varPhi$ = angle subtended in degrees

For a 1° angle subtended and substituted into the above equations, the result is: $(3.14159265 \times 6367449 \times .9998477)/180 = 111116.024$ m or 111.116 km or about 68.9 *statute miles* (sm).

Latitude	Town	Degree	Minute	Second	± 0.0001°
60°	Saint Petersburg	55.80 km	0.930 km	15.50 m	5.58 m
51° 28′ 38″	Greenwich	69.47 km	1.158 km	19.30 m	6.95 m
45°	Bordeaux	78.85 km	1.310 km	21.90 m	7.89 m
30°	New Orleans	96.49 km	1.610 km	26.80 m	9.65 m
0°	Quito	111.3 km	1.855 km	30.92 m	11.13 m

TABLE 11.2   Longitudinal Length Equivalents at Selected Latitudes

Computing Longitudinal Length

Computing longitudinal length is a bit complex as I mentioned above. I used the *haversine formula* to calculate the great circle distance between any two points on the Earth's surface. Great circle distance is often called *as the crow flies*, meaning the shortest distance between two points. The haversine formula was first published by Roger Sinnott in the *Sky & Telescope* magazine in 1984. The formula is actually in three parts, where the first step is to calculate the *a* parameter. Technically, *a* is the square of half of the chord between the two coordinate positions. The second step computes *c*, which is the angular distance expressed in radians. The last step is to compute *d*, which is the linear distance between the two points. The complete haversine formula is shown below:

$$a = sin^2\left(\frac{\Delta\Phi}{2}\right) + cos(\Phi_1) \times cos(\Phi_2) \times sin^2\left(\frac{\Delta\lambda}{2}\right)$$
$$c = 2 \times atan2\left(\sqrt{a}, \sqrt{1-a}\right)$$
$$d = M_R \times c$$

where:

    Φ = Latitude
    λ = Longitude
    M_R = Earth's radius (mean radius = 6371 km)

I am sure that it is possible to calculate this distance manually; however, you have to be very careful in extending the precision of the numbers involved, since some may become very small. I chose to write a Java program to test how the haversine formula actually works with the computer by keeping track of all the tiny numbers involved in the calculations. I did want to mention for those readers with some math background that the two-parameter *atan2* function is used in the second step to maintain the correct sign, which is normally lost when using the single parameter *atan* trignometric function.

I set up a simple test to compare the performance of the haversine formula in determining the path length with Google Earth's straight-path feature. I arbitrarily choose the following coordinates for this test:

    position 1—53°N 1°W
    position 2—52°N 0°W

You should have immediately realized that since one of the positions was on the Prime Meridian, the test must be located in England. Figure 11.11 is a screenshot of Google Earth with the path clearly delineated and the path length shown in the Dialog Box as 130.35 km.

The Java test class named *DistanceDemo.java* was created using the Eclipse *integrated development environment* (IDE) and is shown below:

```java
package distance;
import java.util.*;

public class DistanceDemo {
  final static int MR = 6371; // Earth's radius in km
  static Scanner console = new Scanner(System.in);

  public static void main(String[] args) {
  double lat1, lat2, lon1, lon2, dLat, dLon, a, c, d;
```

```
System.out.println("Enter lat1 = ");
lat1 = console.nextDouble();
lat1 = Math.toRadians(lat1); //all angles must be in radians
System.out.println("Enter lon1 = ");
lon1 = console.nextDouble();
lon1 = Math.toRadians(lon1);
System.out.println("Enter lat2 = ");
lat2 = console.nextDouble();
lat2 = Math.toRadians(lat2);
System.out.println("Enter lon2 = ");
lon2 = console.nextDouble();
lon2 = Math.toRadians(lon2);

dLat = lat2 - lat1;
dLon = lon2 - lon1;

a = Math.sin(dLat/2)*Math.sin(dLat/2)
+
Math.sin(dLon/2)*Math.sin(dLon/2)*Math.cos(lat1)
*Math.cos(lat2); //haversine formula
c = 2*Math.atan2(Math.sqrt(a),Math.sqrt(1-a)); //angular distance
d = MR*c; // linear distance

System.out.println("Distance = " + d + " km");
    }
}
```

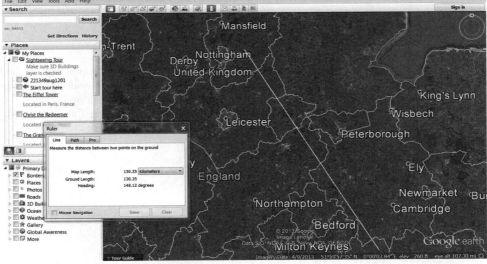

Figure 11.11   Google Earth display showing the test path.

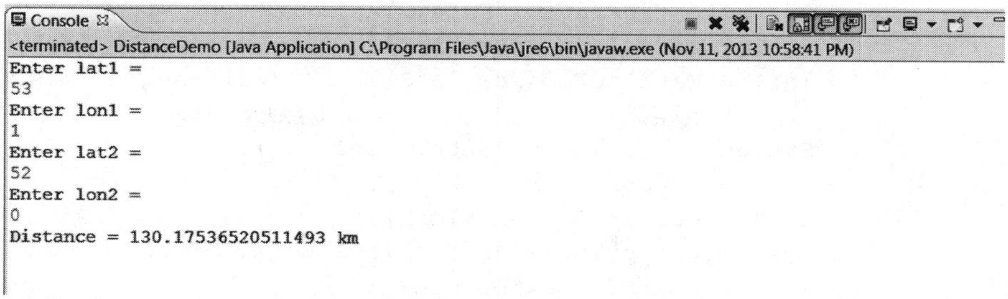

FIGURE 11.12   Eclipse IDE console output screenshot for the DistanceDemo program.

Figure 11.12 is a screenshot of the output from the Eclipse IDE console after the program was run. You can see the coordinates entered as well as the calculated distance of 130.175 km.

I actually believe the results from the Java program are more accurate than the results from Google Earth because the Google Earth path distance depended on how accurately I set the beginning and end points, which is tough to do with the computer's touch pad. In any case, the results differed by about 0.1%, which was enough to convince me that the haversine formula functioned as expected.

There is one more formula that I need to show you in order to complete this navigation discussion. This formula computes the bearing between two coordinates and will be used in conjunction with the electronic compass to guide the quadcopter on the proper path.

Computing Bearing

This formula does not appear to have a formal name, but it is used to calculate the initial bearing between two coordinates by using the great-circle arc as the shortest path distance.

$$\phi = atan2(sin(\Delta\lambda) \times cos(\Phi_2), cos(\Phi_1) \times sin(\Phi_2) - sin(\Phi_1) \times cos(\Phi_2) \times cos(\Delta\lambda))$$

where:
   $\Phi = Latitude$
   $\lambda = Longitude$
   $\phi = Bearing$

This formula, while not nearly as complex as the haversine, will still be demonstrated using a Java program that I aptly named BearingDemo.java. The program code is shown below:

```
package distance;
import java.util.Scanner;

public class BearingDemo {
    static Scanner console = new Scanner(System.in);

    public static void main(String[] args) {
        double lat1, lat2, lon1, lon2, dLat, dLon, phi; // phi is
                                                the forward bearing
```

```
System.out.println("Enter lat1 = ");
lat1 = console.nextDouble();
lat1 = Math.toRadians(lat1);      //all angles must be in
                                        radians
System.out.println("Enter lon1 = ");
lon1 = console.nextDouble();
lon1 = Math.toRadians(lon1);
System.out.println("Enter lat2 = ");
lat2 = console.nextDouble();
lat2 = Math.toRadians(lat2);
System.out.println("Enter lon2 = ");
lon2 = console.nextDouble();
lon2 = Math.toRadians(lon2);

dLat = lat2 - lat1;
dLon = lon2 - lon1;

phi = Math.atan2(Math.sin(dLon)*Math.cos(lat2),
        Math.cos(lat1)*Math.sin(lat2)- Math.sin(lat1)*
        Math.cos(lat2)*Math.cos(dLon));

System.out.println("Bearing = " + phi + " radians");

phi = (phi/Math.PI)*180;          // convert to degrees

System.out.println("Bearing = " + phi + " degrees");
    }
}
```

Figure 11.13 is a screenshot of the output from the Eclipse console after the BearingDemo program was run with the same coordinates as those used in the DistanceDemo program. I measured the bearing with a protractor on a printed copy of Figure 11.11 and estimated the true bearing at 148°, which matches the calculated bearing. Ignore the minus sign on the bearing value, since it is simply a result of the *atan2* function. The two Java programs

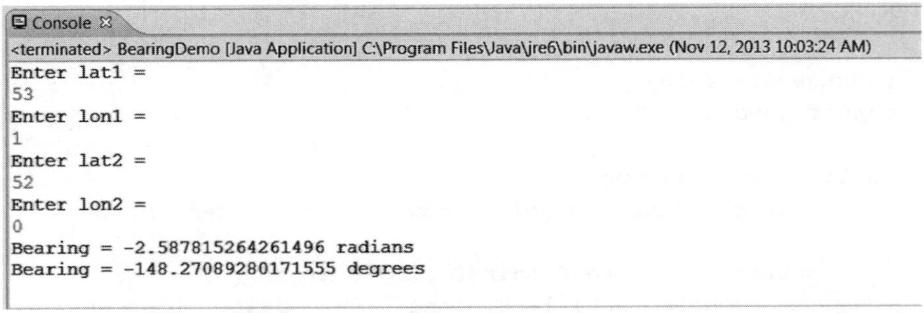

```
□ Console ☒
<terminated> BearingDemo [Java Application] C:\Program Files\Java\jre6\bin\javaw.exe (Nov 12, 2013 10:03:24 AM)
Enter lat1 =
53
Enter lon1 =
1
Enter lat2 =
52
Enter lon2 =
0
Bearing = -2.587815264261496 radians
Bearing = -148.27089280171555 degrees
```

FIGURE 11.13   Eclipse IDE console output screenshot for the BearingDemo program.

used above are available on the book's companion website www.mhprofessional.com/quadcopter.

One more item that must be discussed regarding the bearing is the difference between true bearing and magnetic bearing. True bearings are always taken with respect to the true North (or South) Pole. This is normally the vertical or straight up and down direction on most maps. Magnetic bearings are taken with respect to the magnetic poles, which deviate from the true poles. Called *magnetic deviation*, its value is dependent on where you are taking a compass reading on the earth's surface. In my locale, the deviation is approximately 17° W, which means that I must add 17° to a true bearing in order to determine the equivalent magnetic bearing. For the test location that I used for the program demonstrations, the magnetic deviation was approximately 7° W, thus making the magnetic bearing between the two coordinate positions 155° instead of 148° for the true bearing. There are also magnetic deviations that are classified as *east* (E); these must be subtracted from the true bearing to arrive at the correct magnetic bearing. An old memory aid that pilots and navigators use to help remember whether to add or subtract is the following:

"East is least and West is best" (subtract for East; add for West)

There is also another compass compensation that usually has to be accounted for in non-electronic compasses. This is magnetic declination, or the angle between the Earth's magnetic lines of force and the measuring compass. Fortunately, we do not need to be concerned with this compensation, as it is done automatically by the electronic compass module.

This concludes the navigation fundamentals discussion. You should now have sufficient knowledge regarding how the quadcopter can be guided between geographic coordinate points.

Return-to-Home Flight Scenario

The discussion in this section is based on my vision of how the VFCS functions in a return-to-home scenario. There are commercial flight-control systems that incorporate this type of operation. How the system's designers implement this function is usually proprietary, and thus, not available for analysis. I am guessing that some may use a simple type of dead-reckoning system in which the quadcopter records the angles turned and the time of flight while not turning, and then simply "plays" back these flight motions to return to base. This type of dead reckoning may be perfectly adequate for close-in operations in which distances are short and there is little to no wind to push the quadcopter off its commanded course. The system I envision is much more robust and could easily counter crosswinds and much longer path lengths. It is also suited for autonomous operations that I discuss later in this chapter.

A return-to-home flight command first requires that the coordinates of your initial launch place be stored in the flight controller. This could be done automatically when the quadcopter is first powered on or through a sequence of R/C transmitter commands that the flight controller is programmed to respond to within a predetermined time interval. It could even be implemented by pressing a dedicated button on the quadcopter. Whatever the means, the quadcopter needs to know the start point or home coordinates.

The quadcopter is then flown through whatever flight operations are desired until it receives the command to return home. Sending this command can be done in a fashion similar to the one previously described for storing home coordinates except, of course, for pressing a button on the quadcopter. Let us assume a dedicated R/C channel is used to initiate this operation. This approach is probably the most reliable, although you may not have the luxury of access to an uncommitted channel, especially if you are using an R/C system with six or fewer channels.

1. The first step in returning home is to compute both the path distance and the bearing from the quadcopter's current position to the already stored home position. The bearing is really the most important computation of the two parameters. Path length is nice to know and will be used to slow down the quadcopter; however, you can always manually slow down the quadcopter as it zooms over your home position, as long as it is on the right course.
2. The second step is to command a yaw until the quadcopter is pointed to the correct computed magnetic bearing. Remember that the true bearing is done by the formula, and the appropriate magnetic variation must be added or subtracted to arrive at the correct magnetic bearing.
3. The third step is to command the quadcopter to proceed in the forward direction at a reasonable speed. The path length should be continually recomputed until it reaches a predetermined limit near zero.
4. The fourth step is to slow down and stop the forward motion once the quadcopter is within a reasonable distance of the home coordinates. I would suggest $+/-$ 50 m (54.7 yd) as a good selection.
5. The fifth and final step would be to slowly reduce the throttle-power setting until the hovering quadcopter touches down. This could be done either automatically or under manual control.

The five-step process shown above assumes the existence of little to no crosswind that would push the quadcopter off course. The five-step process could be altered a bit to account for the crosswinds. The bearing would have to be continually recomputed during flight by using the current position. Some minor aileron commands would then be needed to slightly turn the quadcopter to the new home bearing. I would definitely not try to yaw the quadcopter while it is in straight and level flight.

Swarm or Formation Flying

Most likely, you have already seen online videos of quadcopters flying in formation. If not, I would suggest viewing this video, http://makezine.com/2012/02/01/synchronized-nano-quadrotor-swarm/.

Formation flying has been a hot-topic research item for several years. The research focuses on the swarming behavior of insects as they act in a collective manner. There are important goals that researchers are trying to achieve, in which groups or teams of quadcopters can accomplish tasks simply not possible with a single quadcopter. The *General Robotics, Automation, Sensing, and Perception* (GRASP) Lab at the University of Pennsylvania has been the predominant organization in this field. One of its chief researchers is Professor Vijay Kumar, who gave an excellent 15-minute *Technology, Entertainment, Design* (TED) presentation back in February, 2012 that I strongly urge you to view before proceeding in this chapter. Here is the link http://www.ted.com/talks/vijay_kumar_robots_that_fly_and_cooperate.html. This informative video provides a wealth of information on quadcopter performance and swarming behavior.

At present, two technologies are used to implement swarming behavior: *motion capture video* (MoCap) and *close-proximity detection*. First, I will discuss MoCap technology and then the close-proximity detection technology.

Motion Capture

In the TED video, the MoCap cameras can be seen mounted high on the GRASP Lab walls where they have a clear field of view of the flight area. Each quadcopter also has some reflective material mounted on it to provide a good light target for the cameras. Each MoCap camera is networked to a host computer that has been programmed to detect the reflective targets and determine all the quadcopter positions in 3D space. It is probable that only one of the quadcopters has been designated as the lead quadcopter and that sensors and decentralized control allow the others to follow it. The host computer would likely control the lead quadcopter through some type of preprogrammed flight path, and all the others would just follow the leader. However, the video http://www.geeky-gadgets.com/quadcopters-use-motion-capture-to-fly-in-formation-video-17-07-2012/ shows a situation in which all the quadcopters are under direct MoCap control as they execute formation flying.

In this case, the MoCap cameras capture all of the quadcopter's reflective markers and send the images to the host computer, which then calculates the position and attitude for each quadcopter. Precise repositioning to within 1 mm is then transmitted to each quadcopter. The control frequency is 100 Hz, which means that position and altitude are recalculated every 10 ms to prevent collisions.

In the GRASP Lab experiments, each quadcopter had a means of determining how close it was to its neighbor(s) and repositioning itself if the distance was too small. I believe that distance was only several centimeters, which is very tight but not as tight as the full MoCap positioning as described above.

Close-Proximity Detection

I will discuss the Parallax Ping sensor as a typical unit that can serve as a close-proximity sensor. The Ping sensor is shown in Figure 11.14. It has the overall dimensions of 1¾ × ¾ × ¾ in, which is a bit on the large size for a sensor.

The Ping sensor uses ultrasonic pulses to determine distances. It works in a way similar to a bat's echo-location behavior. This sensor may be thought of as a subsystem in that it has its own processor that controls the ultrasonic pulses used to measure the distance from the sensor to an obstruction. Figure 11.15 is a block diagram for this sensor. The sensor has a stated specification range of 2 cm (0.8 in) to 3 m (3.3 yd), which I verified by using the test setup described later in this section.

The Ping sensor uses a one-wire signal line by which the host microprocessor (BOE) emits a two-microsecond pulse that triggers the microprocessor on board the sensor to initiate an outgoing acoustic pulse via an ultrasonic transducer.

Figure 11.14 Parallax Ping sensor.

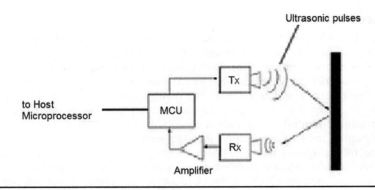

Figure 11.15 Parallax Ping sensor block diagram.

The test code was downloaded from the Parallax product Web page for product number 28015. The test program is named *Ping_Demo_w_PST.spin*. The program continually outputs the distance measured by the sensor in both inches and centimeters. This program also uses a Spin library object named *Ping.spin* that acts as a driver and is listed below:

```
{{
***************************************
*         Ping))) Object V1.2         *
* Author:  Chris Savage & Jeff Martin *
* Copyright (c) 2006 Parallax, Inc.   *
* See end of file for terms of use.   *
* Started: 05-08-2006                 *
***************************************

Interface to Ping))) sensor and measure its ultrasonic travel time.
Measurements can be in units of time or distance. Each method
requires one parameter, Pin, that is the I/O pin that is connected
to the Ping)))'s signal line.
```

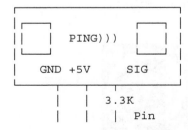

```
                                    Connection To Propeller
                                    Remember PING))) Requires
                                    +5V Power Supply
```

```
--------------------------REVISION HISTORY-----------------------
v1.2 - Updated 06/13/2011 to change SIG resistor from 1K to 3.3K
v1.1 - Updated 03/20/2007 to change SIG resistor from 10K to 1K
}}
```

```
CON
    TO_IN = 73_746                          ' Inches
    TO_CM = 29_034                          ' Centimeters

PUB Ticks(Pin) : Microseconds | cnt1, cnt2      "Return Ping)))'s
                    one-way ultrasonic travel time in microseconds

    outa[Pin]~                              ' Clear I/O Pin
    dira[Pin]~~                             ' Make Pin Output
    outa[Pin]~~                             ' Set I/O Pin
    outa[Pin]~                              ' Clear I/O Pin (> 2 µs
                                              pulse)
    dira[Pin]~                              ' Make I/O Pin Input
    waitpne(0, |< Pin, 0)                   ' Wait For Pin To Go HIGH
    cnt1 := cnt                             ' Store Current Counter Value
    waitpeq(0, |< Pin, 0)                   ' Wait For Pin To Go LOW
    cnt2 := cnt                             ' Store New Counter Value
    Microseconds := (||(cnt1 - cnt2) / (clkfreq / 1_000_000)) >> 1
                                            ' Return Time in µs


PUB Inches(Pin) : Distance              'Measure object distance in
                                          inches
    Distance := Ticks(Pin) * 1_000 / TO_IN   ' Distance In Inches


PUB Centimeters(Pin) : Distance             ' Measure object
                                              distance in centimeters
    Distance := Millimeters(Pin) / 10
' Distance In Centimeters


PUB Millimeters(Pin) : Distance             'Measure object
                                              distance in millimeters
    Distance := Ticks(Pin) * 10_000 / TO_CM      ' Distance In
                                                   Millimeters
```

The following code snippet from the above listing generates the initial pulse that is sent to the sensor. The onboard sensor then sets the signal line high and then listens for the return pulse with another ultrasonic transducer. The echo-return acoustic pulse causes the onboard processor to change the signal line from high to low. The Ping code measures the time interval, in microseconds, that has elapsed while the signal level was going from high to low. The system-clock counter value is stored in the variable cnt1 when the high value is detected and in the variable cnt2 when the sensor changes the signal-line level to low. The difference, cnt2 − cnt1, must then be the elapsed time in the units of system-clock cycles.

```
    outa[Pin]~                              ' Clear I/O Pin (which is
                                              P0 set by the test code object)
    dira[Pin]~~                             ' Make P0 an output pin
    outa[Pin]~~                             ' P0 is set high
```

```
    outa[Pin]~                          ' Make P0 low (This creates
                                          an approximate 2 µs pulse)
    dira[Pin]~                          ' Make P0 an input pin
    waitpne(0, |< Pin, 0)               ' Wait For P0 to go high
    cnt1 := cnt                         ' Now store the current
                                          system clock counter value
                                          into cnt1
    waitpeq(0, |< Pin, 0)               ' Wait For P0 to go low
    cnt2 := cnt                         ' Now store the current
                                          system clock counter
                                          value into cnt2
  Microseconds := (||(cnt1 - cnt2) / (clkfreq / 1_000_000)) >> 1
                                    ' Compute the return time in µs
```

The test code also incorporates a provision to activate two LEDs, depending upon the measured distance. The LED connected to P1 will turn on when the distance is less than 6 inches. The other LED connected to P2 will turn on when the distance exceeds 6 inches. The P1 LED will also turn off when the distance exceeds 6 inches. The test code is shown below:

```
" ****************************************
" * Ping))) Demo with PST & LED's                    *
" * Author: Parallax Staff                               *
" * Started: 06-03-2010                                  *
" ****************************************
{{
Code Description: In this example there are two LED's to indicate a
distance. If the distance is further  than 6 inches than LED 1 will
turn on, and if the distance is closer than 6 inches than LED 2
turns on; while either LED 1 or 2 is on, the alternate LED will be
off. There is a numerical display of the values in the Parallax
Serial Terminal (PST) at 9600 baud (true).
}}

CON
  _clkmode = xtal1 + pll16x
  _xinfreq = 5_000_000

  PING_Pin      = 0                     ' I/O Pin For PING)))
  LED1          = 1                     ' I/O PIN for LED 1
  LED2          = 2                     ' I/O PIN for LED 2

  ON            = 1
  OFF           = 0
  Distlimit     = 6                     ' In inches

VAR
  long  range
```

```
OBJ
  Debug  : "FullDuplexSerial"
  ping   : "ping"

PUB Start
  dira[LED1..LED2]~~
  outa[LED1..LED2]~

  Debug.start(31,30,0,9600)
  waitcnt(clkfreq + cnt)

  repeat                                ' Repeat Forever
    debug.str(string(1,"PING))) Demo ", 13, 13, "Inches = ", 13,
         "Centimeters = ", 13))

    debug.str(string(2,9,2))
    range := ping.Inches(PING_Pin)      ' Get Range In Inches
    debug.dec(range)
    debug.tx(11)

    debug.str(string(2,14,3))
    range := ping.Millimeters(PING_Pin) ' Get Range In
                                        '   Millimeters
    debug.dec(range / 10)               ' Print Whole Part
    debug.tx(".")                       ' Print Decimal Point
    debug.dec(range // 10)              ' Print Fractional
                                        '   Part

    debug.tx(11)

    range := ping.Inches(PING_Pin)      ' Get Range In
                                        '   Inches

    if range < Distlimit                ' Comparing range to a
                                        '   set value of 6 inches

      outa[LED1] := ON                  ' P1 is on
      outa[LED2] := OFF                 ' P2 is off
    elseif range > Distlimit            ' If range is
                                        '   further than 6 inches

      outa[LED1] := OFF                 ' P1 is off
      outa[LED2] := ON                  ' P2 is on
```

Figure 11.16 shows the setup of the test components on the BOE solderless breadboard. The P1 LED is at the top of the breadboard, and P2 is near the bottom. A book was placed 8 in from the Ping sensor to reflect the ultrasonic pulses. A PSerT screenshot shown in Figure 11.17 illustrates the results of the test setup shown in Figure 11.16.

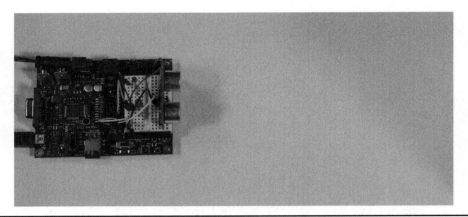

FIGURE **11.16**   Ping sensor test setup.

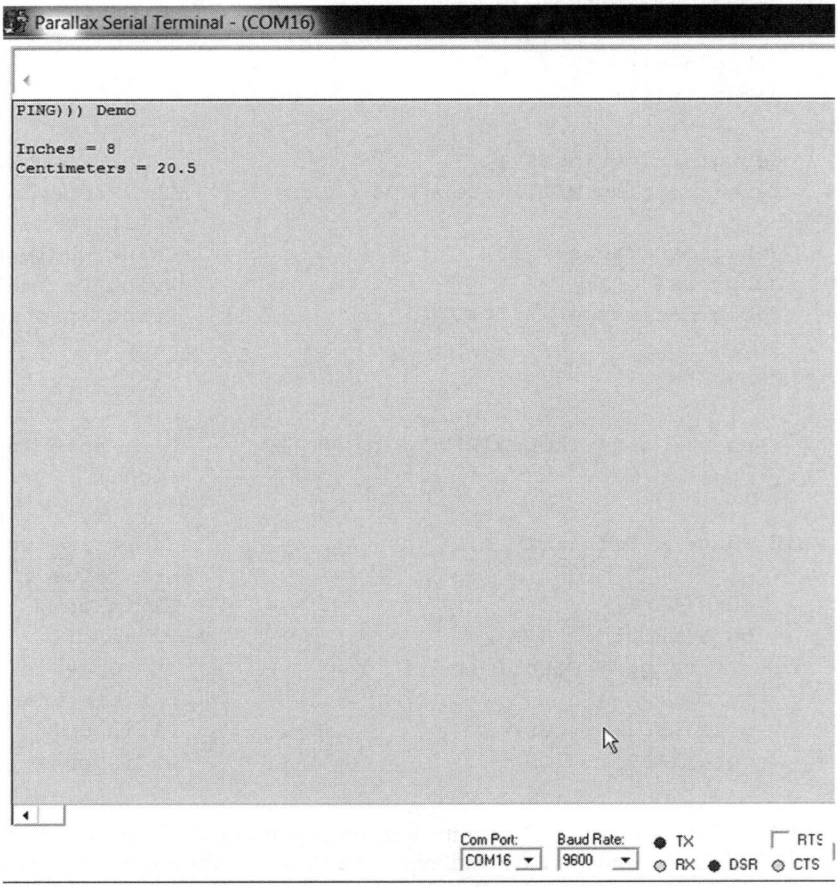

Parallax Serial Terminal - (COM16)

```
PING))) Demo

Inches = 8
Centimeters = 20.5
```

Com Port: COM16   Baud Rate: 9600   ● TX   ☐ RTS
○ RX ● DSR ○ CTS

FIGURE **11.17**   PSerT screenshot for the Ping sensor test.

Near-to-Ground Altitude Measurements

A Ping sensor can also be mounted on the underside of the bottom chassis plate, pointing straight down, which would provide near-to-ground altitude measurements. The maximum Ping sensor range is 3 m (3.3 yd), which is more than enough to provide good above-ground readings for hovering or for approach-to-landing operations. Figure 11.18 shows a Ping sensor that is mounted on the bottom chassis plate and has a clear field of view of the ground.

It could also be used as part of the close-proximity sensor set discussed in the section on formation flying. In such a situation, it would provide vertical clearance measurements between it and any quadcopter flying below it.

Ultrasonic Sensor Concerns

There are some conditions that you should be aware of if you want to use an ultrasonic sensor successfully for close-proximity detection. These conditions are listed and discussed below:

- Wind turbulence
- Propeller acoustic noise
- Electrical noise both conducted and radiated
- External radio interference
- Frame vibration

Wind turbulence is created by the propellers. The only realistic solution to minimize this type of interference is to mount the sensor as far as possible from any propeller.

Propeller acoustic noise adds additional acoustic energy to the sensor, which generally reduces the overall ultrasonic transducer sensitivity. Careful sensor placement reduces this effect along with avoiding a direct structural mount near any motor.

The ultrasonic sensor's electrical power supply should be directly connected to the same power source used by the flight-controller board. In the Elev-8 configuration,

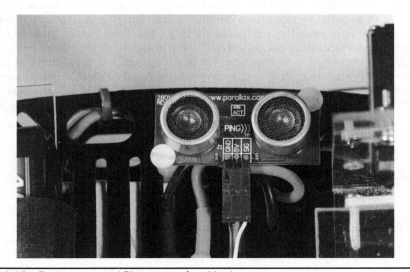

Figure 11.18 Bottom-mounted Ping sensor for altitude measurements.

the HoverflyOPEN controller-board power is supplied through the BEC lines (as I described in Chapter 5). The AR8000 receiver, in turn, is powered from the flight-control board. Using the power-distribution board actually helps reduce the interference by providing a ground-plane-shielded power source for the ESCs, which are primary potential noise sources.

Strong interference may also be present as conducted electrical noise. This type of noise is often eliminated by using a simple *resistor/capacitor* (RC) filter. Figure 11.19 shows an RC filter that can be connected at the sensor power inputs to eliminate any conducted electrical noise interference.

Electrical currents that flow through wires produce an *electromagnetic* (EM) field. Strong currents that also carry noise pulses produce what is known as radiated electrical noise. The ultrasonic-sensor power leads should be twisted to mitigate any possible radiated noise interference. You also might need to use a shielded power cable if the interference is particularly severe. Ground only one end of the shielded cable at the host microprocessor side to stop any possible ground loop current from forming.

There are also several radio transmitters on board the Elev-8 that can generate lower levels of EM interference. Usually they will not be an issue provided you have taken some or all of the control measures already mentioned.

The final interference might come from frame vibrations, which can upset the sensor's normal operation. This type of interference is easily minimized by securing the sensor in a small frame that, in turn, is mounted on the quadcopter's frame with rubber grommets. This is precisely the same type of mounting arrangement that is used to mount the HoverflyOPEN control board on the Elev-8.

Maneuvering the Quadcopter to Maintain Its Formation Position

Maintaining position in a formation is really a matter of providing very slight control inputs to shift the quadcopter's position a few centimeters. Ping sensors mounted at the ends of each quadcopter boom are easily programmed to provide an output on two pins that can signal to the flight controller that the quadcopter is either closer to or farther away from a neighboring quadcopter than it should be. The precise control inputs that are needed will probably be determined by trial and error, since only small shifts in rotor rotation speeds would be required to accomplish centimeter-scale movements. It would make no sense to actually bank or pitch the quadcopter for tiny lateral movements. Massive control inputs, such as banking or pitch changes, would result in excessivly large displacements, which are definitely not required. Instead, just changing the rotation speeds on one or two motors by 100 to 200 r/min may very well accomplish the required shifts. Such precise control will likely need to be repeated 10 to 20 times per second to maintain accurate positioning.

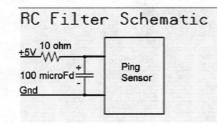

FIGURE 11.19 RC noise filter for conducted electrical noise.

Figure 11.20   The MaxBotix HRLV-Max Sonar®– EZ1™, model MB1013.

Other Close-Proximity Sensors

Several other manufacturers provide ultrasonic sensors that will function quite well for close-proximity detection. Figure 11.20 shows the MaxBotix HRLV-Max Sonar®–EZ1™, model MB1013, which is a very compact sensor with overall dimensions of approximately ¾ × ¾ × ¾ in.

This versatile sensor has many more operational functions than the Ping sensor. The following summary in Table 11.3 was extracted from the MaxBotix data sheet, to provide you with some additional background on this sensor type. The pin connections for this sensor are shown in Figure 11.21.

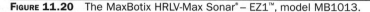

Pin No.	Name	Description
1	Temperature sensor	Connect external temperature sensor to improve overall accuracy.
2	Pulse-width output	This pin outputs a pulse-width representation of the distance with a scale factor of 1 microsiemens (μS) per mm. Output range is 300 μS for 300 mm (11.81 in) to 5000 μS for 5000 mm (196.85 in).
3	Analog voltage output	This pin outputs an analog voltage-scaled representation of the distance with a scale factor of (V_{cc}/5120) per 1 mm. The distance is output with a 5-mm resolution.
4	Ranging start/stop	If this pin is left unconnected, or held high, the sensor will continually measure and output the range data. If held low, the HRLV-MaxSonar-EZ will stop ranging.
5	Serial output	The serial output is RS232 format (0 to V_{cc}) with a 1-mm resolution. If TTL output is desired, solder the TTL jumper pads on the back side of the PCB, as shown in Figure 11.20.
6	V_{cc}	The sensor operates on voltages from 2.5 V to 5.5 V DC.
7	GND	This is the sensor ground pin.

Table 11.3   MaxBotix Ultrasonic Pinout Descriptions

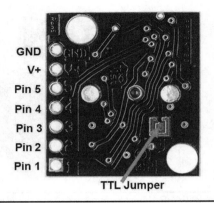

GND
V+
Pin 5
Pin 4
Pin 3
Pin 2
Pin 1

TTL Jumper

FIGURE 11.21   Pin descriptions for the MaxBotix ultrasonic sensor.

Real-Time Range Data—When pin 4 is low and then brought high, the sensor will operate in real time, and the first reading that is output will be the range measured from this first commanded range reading. When the sensor tracks that the RX pin is low after each range reading and then the RX pin is brought high, unfiltered real-time range information can be obtained as quickly as every 100 ms.

Filtered Range Data—When pin 4 is left high, the sensor will continue to range every 100 ms, but the output will pass through a 2-Hz filter, through which the sensor will output the range, based on recent range information.

Serial Output Data—The serial output is an ASCII capital R, followed by four ASCII digit characters representing the range in millimeters, followed by a carriage return (ASCII 13). The maximum distance reported is 5000 mm. The serial output is the most accurate of the range outputs. Serial data sent is 9600 baud, with 8 data bits, no parity, and one stop bit.

One important constraint that you must be aware of is the Sensor Minimum Distance or No Sensor Dead Zone. The sensor minimum reported distance is 30 cm (11.8 in). However, the HRLV-MaxSonar-EZ1 will range and report targets to within 1 mm (0.04 in) of the front sensor face; however, they will be reported as at a 30-cm (11.8 inches) range.

Another close-proximity sensor that you might consider using is one based on invisible light pulses. This sensor is the Sharp GP2D12, which is an *infrared* (IR) light-ranging sensor. It is shown in Figure 11.22.

FIGURE 11.22   Sharp GP2D12 IR range sensor.

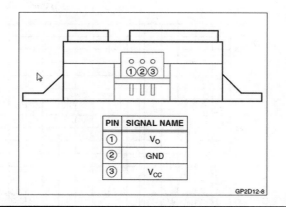

PIN	SIGNAL NAME
①	V_O
②	GND
③	V_{CC}

GP2D12-8

FIGURE 11.23   Sharp GP2D12 sensor pinout.

The IR light pulses used by this sensor are similar to the IR pulses used in common TV remote controls and are rather impervious to ambient light conditions. The pinout for this sensor is shown in Figure 11.23 and it has only three connections, as was the case for the Ping sensor. The V_{CC} supply range is from 4.5 to 5.5 V, and the output is an analog voltage that is directly proportional to the target range. Figure 11. 24 shows a graph of the V_O pin voltage versus the target range.

This type of output means that you must use an *analog-to-digital converter* (ADC) to acquire the numerical range value. The range precision will also be dependent on the number of ADC bits. The Parallax Propeller chip uses 10-bit sigma-delta ADCs, which means a 1024-bit resolution over the input voltage range. Figure 11.23 indicates a maximum output of 2.6 V at 10 cm (3.94 in), which incidentally, is also the minimum range the sensor will detect. I would probably use the V_{CC} as an ADC reference voltage, since it is readily available and encompasses the maximum expected input voltage.

The precision calculations would be as follows:

$$\text{Max possible input voltage}/2^{10} = 5/1024$$
$$= 0.004883 \text{ V per count or}$$
$$= 4.883 \text{ mV per count}$$

The maximum range is 80 cm (31.5 in), which generates a 0.4 V output. Therefore, the total voltage change for the specification range of 10 to 80 cm must be:

Voltage at 10 cm	2.6
Voltage at 80 cm	-0.4
Interval voltage	2.2

Next divide 2.2 V by 70 to arrive at volts per centimeter (Note: I am assuming linearity, which is not actually the case, but it will have to suffice without overly complicating things.)

$$2.2/70 = .03143 \text{ V per cm or}$$
$$= 31.43 \text{ mV per cm}$$

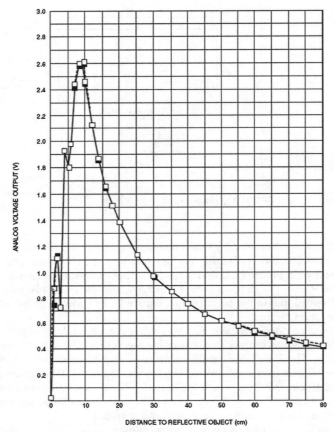

FIGURE 11.24   Analog voltage out versus target range.

Now using the 4.883 mV per count calculated above, it is easy to see that the ultimate precision is:

(31.43 mV per cm)/(4.883 mV per count) = 6.44 counts per cm or
rounding to 6 counts per cm.

This means each centimeter in the interval can be resolved to approximately one sixth of a centimeter or about 16 mm (0.63 in), which should be sufficient for most close-proximity operations.

The above calculations are fine, but you are likely puzzled as to the actual numbers that you could expect from the ADC. Three values are shown below that represent the minimum, midpoint, and maximum ranges:

Minimum $(2.6/5.0) \times 1024 = 532$
Midpoint $(0.68/5.0) \times 1024 = 139$
Maximum $(0.40/5.0) \times 1024 = 82$

I always like to do a sanity check on my calculations, which follows:

$$\begin{aligned}
\text{Count at 10 cm} &= 532 \\
\text{Count at 80 cm} &= -82 \\
\hline
\text{Count difference} &= 450 \text{ which represents the 70 cm interval}
\end{aligned}$$

450 counts/6.44 counts per cm $\cong$ 70 cm   Checks out!

I did mention in the above calculations that I assumed that analog voltage versus distance curve was linear or a straight line, which it is obviously not. If you desire absolute precision, you would have to develop either a lookup table or an analytic equation that modeled the curve. The latter is actually not hard to do by using the Microsoft Excel feature that automatically creates best-fit equations from a data set, but that is best left for another time.

One more special type of proximity sensor I would like to discuss is a LIDAR. The term *LIDAR* is a combination word made from the words light and radar. It uses an IR laser beam to detect, and many times, map out distant objects. Until recently, LIDAR sensor systems were bulky, consumed substantial power, and were very expensive. But that has changed recently, to the point where very capable systems are now available that can be mounted on a quadcopter. They are also relatively inexpensive. LIDAR is able to range distant objects at or beyond 3 km (1.86 mi) because it uses high power and IR laser pulses of very short duration. The reflected pulses are detected by sensitive, optical-photo receivers, and the distances are computed by precisely the same method used by traditional radar systems.

Figure 11.25 shows a LIDAR kit, model ERC-2KIT, that is sold by Electro-Optic Devices. The kit contains a single-board ranging controller as well as the transmitter and the receiver boards, as shown in the figure. It does not come with a laser diode, which must be purchased

Single board ranging controller

Low-noise Si-PIN photodiode optical receiver

MOSFET-based pulsed laser diode driver w/ filtering

FIGURE 11.25   LIDAR system model ERC-2KIT, from Electro-Optic Devices.

Figure 11.26   A LIDAR-pulse diode.

separately. Diode selection depends upon the intended LIDAR application because diodes are available in a range of both wavelength and power capacity. Common wavelengths are 850 and 905 nanometers (nm), both of which are in the IR range. Power ratings can vary from a low of 3 W to a high of 75 W, which is a very powerful laser that can cause severe eye injury if not used carefully.

Figure 11.26 shows the OSRAM SPL PL85 LIDAR-capable laser-pulse diode that is rated at 10 W and capable of ranging up to about 1 km (0.62 mi). LIDAR pulses are very short in time duration, typically some tens of nanoseconds. However, the current pulse can easily exceed 20 A.

Electro-Optic Devices provides a test and control board for their LIDAR ERC-2KIT, which is very useful for development purposes. Electro-Optic Devices calls this board the BASIC Programmable Laser Ranging Host Module, model EHO-1A, and it is shown in Figure 11.27.

This board uses a Parallax Basic Stamp II (BS2) as a controller, which fits nicely into the whole Parallax controller discussion that has been ongoing in this book. The BS2 uses a derivative of the BASIC language named PBASIC to implement its microcontroller functions. BASIC, as most of you already know, is procedural and not object oriented as is the Spin language used for the Parallax Propeller microcontroller. Nonetheless, it is more than adequate for this application, and it is really very easy to program in BS2 BASIC. The BS2 also uses The Basic Stamp Editor, which is a different *integrated development environment* (IDE) from the Propeller chip IDE. It is freely available to download from the Parallax website.

The PBASIC instructions in the following section are excerpted from the Electro-Optic Devices's EHO-1A manual to illustrate how to display data on the LCD screen and how to read from and write to the peripheral modules.

(LCD removed for illustration)

Figure 11.27   BASIC Programmable Laser Ranging Host Module, model EHO-1A, from Electro-Optic Devices.

Writing to the EHO-1A LCD Screen

The LCD display must be written to serially from the BS2. The BS2 instruction SHIFTOUT accomplishes this task. There are four LCD-oriented subroutines available in each example program. They are:

1. DSP_INIT—Initializes the LCD Display
2. DSP_TEXT—Sends a text string to the LCD
3. DSP_CLR—Clears the LCD
4. DSP_DATA—Displays binary data on the LCD (up to 4 digits)

At the top of each example program is a section for EEPROM DATA. ASCII string data is stored in this location to be written to the LCD by the DSP_TEXT subroutine. The LCD string format is as follows:

```
LABEL DATA L#P#,LENGTH, "STRING INFO"
```

The LABEL can be any valid name for the string. DATA indicates to the tokenizer that this information will be stored in EEPROM. L#P# is the line number (1 or 2) and position number (1–16) where the first character of text is to be located in the LCD. LENGTH is the number of characters that follow inside the quotation marks. For the above example, this program line might look like this:

```
T_STRG DATA L2P1,11, "STRING INFO"
```

When the DSP_TEXT subroutine is called after the message pointer MSG is set to the string's label: MSG = T_STRG, the text message "STRING INFO" is displayed at the first position of line 2 on the LCD. Similarly, BCD data can be written to the LCD using the DSP_DATA subroutine. A binary number to be displayed in the range of 0–9999 must be stored in VALUE. The LCD location pointer LOC must be set to the location of the first digit of a four-digit result. Before calling the DSP_DATA subroutine, a selection must be made between the two decimal point formats for the four-digit display (NNNN or NNN.N). Two examples follow:

1. To display the number 1234 beginning at character position 10 in line 1 of the LCD:

```
VALUE = 1234
LOC = L1P10
D_FLAG = XXXX
GOSUB DSP_DATA
```

2. To display the number 456.7 beginning at character position 3 in line 2 of the LCD:

```
VALUE = 4567
LOC = L2P3
D_FLAG = XXX_X
GOSUB DSP_DATA
```

Communicating with Peripheral Modules

The example programs each have slight differences in the data acquisition subroutines and should be examined closely before writing your own. The ECH-4 is especially unique, since it uses a single bidirectional serial bus. The basic procedure is described in the following section.

Writing to a Peripheral Module

Set the COMMAND variable to the command or data byte to be written to the module, for example:

```
COMMAND = $00
```

Next, bring the chronometer select signal $\overline{CS}$ low to enable communication with the module. Now use the BS2 SHIFTOUT instruction to send the byte to the module,

```
SHIFTOUT HDO, HCLK, MSBFIRST, [COMMAND]
```

Now bring $\overline{CS}$ back high to disable communication and acknowledge the end of a write.

Reading from a Peripheral Module

Each module has its own manner of indicating that data is available to be read. See the individual examples for more information. Generally, the read is performed similarly to the write. Bring the chronometer select signal ($\overline{CS}$) low to enable communication with the module. Now use the BS2 SHIFTIN instruction to read the byte from the module:

```
SHIFTIN HDI, HCLK, MSBPRE, [DATABYTE]
```

Now bring $\overline{CS}$ back high to disable communication and acknowledge the end of a write. The read information is now stored in the variable DATABYTE.

Autonomous Behavior

Autonomous behavior happens when the quadcopter performs tasks without direct human operator control. A common autonomous task might be to fly a preset path, which could even include taking off and landing without any human intervention. Such a task would necessarily have to have predetermined coordinates already programmed into the flight-controller's memory. These coordinates are normally called *waypoints* and are just a series of latitude and longitude coordinate sets that the quadcopter will fly to in a preset sequence. Of course, flying a preset path without any other function would be fairly meaningless beyond giving you the satisfaction of being able to do it. Taking video or periodic still photographs while flying the path would be a much more meaningful experience and would illustrate the versatility of the quadcopter. Photographic aerial damage assessment after a natural disaster would be a good fit for an autonomous quadcopter mission.

Creating a virtual map of an indoor environment is another interesting task that is currently being developed by a number of organizations. The quadcopter would be equipped with a type of LIDAR as discussed above. The LIDAR would usually be mounted on an automated pan-and-tilt mechanism. A typical and rather inexpensive mount is shown in Figure 11.28.

FIGURE 11.28   Pan-and-tilt mechanism for LIDAR mapping.

Using the ERC-2KIT requires only that the lightweight transmitter and receiver boards be mounted with a pair of signal cables connected to the ranging controller. In a video shown in Professor Kumar's TED presentation, you can see mapping being done by a quadcopter within a building. Incorporating *artificial intelligence* (AI) within the flight controller helps the quadcopter avoid obstacles and keeps it from being trapped in a room. This important topic is discussed further in the next section because it is a vital component to the successful completion of an indoor mapping task.

Artificial Intelligence

Artificial intelligence is an extremely interesting topic that I have studied for a number of years. I will not present a full AI discussion but instead will focus on the essential ideas that are directly applicable to quadcopter operations in a confined space. The essential goal for this specific AI application is to equip the quadcopter controller with sufficient "reasoning" capability so that it can autonomously avoid obstacles.

Researchers have already determined that the AI methods of *Fuzzy Logic* (FL) and, more specifically, *Fuzzy Logic Control* (FLC) are particularly well suited for autonomous robotic operations. Let me start by stating that there is nothing "fuzzy" or "confused" about this AI field of study; it is simply a name applied to reflect that it incorporates decision ranges in lieu of traditional, discrete, decision points of yes/no, true/false, equal/not equal, and so on. I need to show you some basic FL concepts before proceeding any further.

Some Basic FL Concepts

Professor Lufti Zadeh invented FL in 1973. He applied set theory to traditional control theory in such a way as to allow imprecise set membership for controlling purposes instead of using normal, precise, numerical values, as was the case before FL. This impreciseness allows noisy and somewhat varying control inputs to be accommodated in ways that were not previously possible.

FL relies on the propositional logic principle, *Modus Ponens* (MP), which translates to "the way that affirms by affirming." An equivalent logical statement is:

$$\text{IF } X \; AND \; Y \text{ THEN } Z$$

where *X AND Y* is called the antecedent and *Z* is the consequent.

Next, I will use a simple room-temperature-control example to demonstrate the fundamental parts that make up an FLC solution. Normally, you might have the following control algorithm in place to control an indoor room:

IF Room Temperature $<= 60°$ F THEN Heating System $=$ On

This reflects precise measurements and a definitive control action. It is also easily implemented by a "dumb" thermostat. Transforming the above control statement into an FL type statement might lead to:

IF (Room too cool) THEN (Add heat to room)

You should readily perceive the input statement's impreciseness, yet there is a certain degree of preciseness in the outcome or output action. However, do not make the mistake of thinking that FL does not use numerical values; it does, but they are derived from a set of values.

Room temperatures may be grouped or categorized into regions with the following descriptors:

- Cold
- Cool
- Normal
- Warm
- Hot

If you were to survey a random group of people, you would quickly find that one person's idea of warm might be another person's idea of hot, and so forth. It quickly becomes evident that some type of membership value must be assigned to different temperatures in the different regions. In addition, a graph of temperature-versus-membership value could take on different forms, depending upon how the survey was taken. Triangular and trapezoidal are the two graph shapes that are normally used with membership functions and that vastly simplify FL calculations. Membership values range from 0 to 100%, where 0 indicates no set members are present, while 100% shows all set members are within the region. By set members, I mean the discrete temperatures on the horizontal axis. Figure 11.29 shows the five membership functions for the five temperature regions.

The use of these shapes for membership functions avoids arbitrary thresholds that would unnecessarily complicate passing from one region to the next. It is entirely possible, and in fact desirable, that a given temperature have membership in two regions

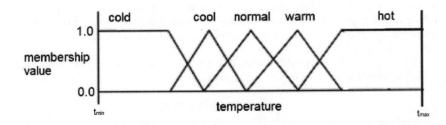

FIGURE 11.29 Temperature membership functions.

simultaneously. Not every temperature point has to have dual or even triple membership, but temperatures on the region's edges should be so assigned except for the extremis ones associated with the very cold and the very hot trapezoidal regions, for example, t_{min} and t_{max} of Figure 11.29. This format allows the input variable to gradually lose membership in one region, while increasing membership in the neighboring region. The translation of a specific temperature to a membership value in a selected region is known as *fuzzification*. The complete set of membership values for the input temperature is also referred to as a *fuzzy set*.

The combination of fuzzy sets and Modus Pons relationships are the needed inputs to a rule set that decides what action to take for the control inputs. FLC may be analyzed as three stages:

1. *Input*—This stage takes sensor input and applies it to the membership function to generate the corresponding membership values. Values from different sensors may also be combined for a composite membership value.
2. *Processing*—Takes the input values and applies all the appropriate rules to eventually create an output that goes to the next stage.
3. *Output*—Converts the processing result to a specific control action. This stage is also called *defuzzification*.

The processing stage may have dozens of rules, all in the form of IF / THEN statements. For example:

IF (temperature is cold) THEN (heater is high)

The antecedent IF portion holds the "truth" input value that the temperature is "cold," which triggers a "truth" result in the heater-output fuzzy set that its value should be "high." This result, along with any other valid rule outputs, is eventually combined in the output stage for a discrete and specific control action: so called defuzzification. You should also note that, as a given rule, the stronger the truth-value is for an input, the more likely it will result in a stronger truth-value for the output. The resultant control action may not be the one expected, however, since control outputs are derived from more than one rule. For instance, in the example of the room heating and cooling system, if a fan were used, it is entirely possible that the fan's speed might be increased, depending upon the rule set and whether or not the heater is set on high.

The AI core of FL lies in the construction of the rule set, which basically encapsulates the knowledge of problem-domain experts in crafting all the rules. Typically, subject matter experts (SMEs) would be asked a series of questions, such as "if so and so happened, what would your response be?" These questions and the SME answers would then constitute the rules set as a series of IF/THEN statements. The usefulness of an FLC solution is totally dependent upon the quality of the SME input.

This concludes my FL basics introduction, and I will now return to how FLC is used with quadcopters.

Quadcopter FLC Applications

There are two common types of FLCs:

1. Mamdani
2. Sugeno

Mamdani types are the standard FLCs, in which there are membership shapes for both their input and output. Defuzzification of the output fuzzy variables is done by using a center-of-gravity, or centroid, method. *Sugeno* types are simplified FLCs, in which only inputs have membership shapes. Defuzzification is done by the simpler weighted average method. I will be discussing only the Mamdani FLC because that seems to be the most popular approach for quadcopter FLC.

FLC commonly uses the term error (*E*), which is the actual output minus the desired output. I will use *Z* as the variable for the error input. It is also very common to use the rate of change of the error variable as an input, which I will show as *dE* where the *d* represents the time derivative of *E*. Finally, accumulated error should be accounted for. It will be represented by *iE* where the *i* represents the error integrated (summed) over a time interval. If all this looks vaguely familiar, I will refer you back to Chapter 3 where the *proportional, integral, and derivative* (PID) controller was introduced. FLC controllers function as PID controllers with some additional pre- and postprocessing to account for the FL components.

The input variables *Z*, *dE*, and *iZ* also need to be multiplied by their gains, *GE*, *GDE*, and *GIE* respectively. The output variable will be designated *Z*, and it also has a gain of *GZ*.

Matlab®

Matlab® is a powerful, scientific-modeling and math computational system that will be the basis for the rest of this discussion. Matlab® has a variety of toolkits that expand its capabilities, including a Fuzzy Logic Toolkit. This toolkit has the following characteristics:

- Mamdani inference.
- Triangular central membership function with the rest as trapezoidal shapes.
- The FL rules set will follow this form:
  if (Ez is E) and (dEz is DE) and (iEz is IE) then (Z is Zz), where E, DE, IE, and Zz are fuzzy sets.
- Defuzzification method is centroid or center of gravity.
- AND operator implemented as the minimum.
- Implication is a minimum function.
- Tuning of input and output gains will be done by trial and error.

Figure 11.30 is a block diagram of a Matlab® quadcopter control system. There are four FL controllers shown in the diagram: Z, roll, pitch, and yaw. The Z controller acts as a kind of master because it controls altitude or height above the ground. Obviously, if the quadcopter is not above the ground, it is not flying; and the three other controller actions are moot. Each FL controller takes the four inputs that are the same primary ones first mentioned in Chapter 3:

1. Throttle
2. Elevator
3. Aileron
4. Rudder

Each FL controller also has four outputs, one for each of the quadcopter motors. You may be able to see the membership functions drawn as three generic-triangle shapes in each of the FL controller blocks. They simply symbolize that each block constitutes an input stage. The Aggregation block is a combination processing and output stage that holds all the fuzzy set rules as well as the logic to combine all four FL controller outputs.

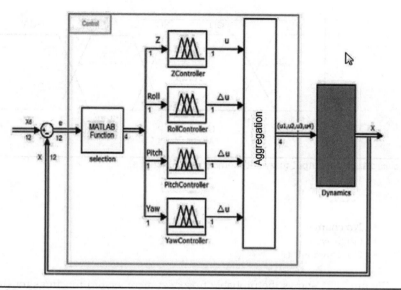

FIGURE 11.30   Matlab® quadcopter control-system block diagram.

The Z controller always has equal power supplied to each motor so that only vertical travel is allowed. If F_z is one output, then $4F_z$ must be supplied to all four motors. This power level must always be maintained if the quadcopter is to remain at a commanded altitude. Other controllers can require certain motors to speed up and others to slow down in order to achieve a yaw, pitch, or roll, but the net overall power will always result in a net of $4F_z$. It is entirely possible that the sets of fuzzy rules will limit or prohibit certain combinations of control inputs, since they would be physically impossible to perform.

The Z controller is probably the easiest to understand because it controls motion along only one axis. Let's say that the proportional control input (E) is an error signal that is the difference between the commanded altitude and the actual altitude. Also present will be the derivative (dE) and integral (iE) inputs that makeup the totality of the PID control system. The likely proportional input fuzzy sets might be:

- Go up
- Hover
- Go down

The derivative and integral input fuzzy sets might be:

- Negative
- Equal
- Positive

The output fuzzy set, assuming a Mamdani setup, might be:

- Go up a lot
- Go up

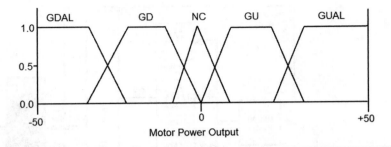

FIGURE 11.31 Z-output fuzzy set membership functions.

- No change
- Go down
- Go down a lot

Figure 11.31 shows the probable fuzzy set membership functions for the Z parameter using the listed control actions. Notice that four out of five membership shapes are trapezoidal and only the No Change is sharply triangular, thus reflecting the expert opinions that most often some control action is needed to maintain altitude.

Table 11.4 shows the probable rules outputs given all nine combinations for the three Z output variables Z, dZ, and iZ. Remember that these outcomes are suggested by SMEs given the stated conditions for the membership variables. Sometimes this rule table is referred to as an inference table, reflecting the MP background.

dZ-iZ	Z	Up	No Change	Down
Negative	Negative	GDAL	GD	NC
Negative	Equal	GDAL	GD	NC
Negative	Positive	GDAL	GD	NC
Equal	Negative	GD	GU	GU
Equal	Equal	GD	NC	GU
Equal	Positive	GD	GU	GU
Positive	Negative	GU	GU	GUAL
Positive	Equal	GU	GU	GUAL
Positive	Positive	GU	GU	GUAL

Where:

GUAL = Go up a lot
GU = Go up
NC = No change
GD = Go down
GDAL = Go down a lot

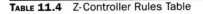

TABLE 11.4 Z-Controller Rules Table

Of course, the relative results of the error terms depend very much upon the gains associated with their PID inputs. A larger gain will add more weight to a specific input, which could cause somewhat of a biased operational result. Too much gain can, and often will, result in unstable or oscillatory behavior that makes the quadcopter unflyable.

ViewPort™ Fuzzy Logic Functions

ViewPort™ is a software development tool for the Propeller chip that was developed and marketed by myDancebot.com, a Parallax partner. ViewPort™ contains a fuzzy logic view as part of the ViewPort™ Development Studio software suite. You can incorporate FL objects into your programs with this tool. It will greatly enhance your ability to create quadcopter FLCs based on the Propeller chip and the Spin language.

The following discussion has excerpts from the ViewPort™ manual to help clarify how FL could be incorporated into Spin programs.

> *ViewPort comes with a Graphical Control Panel view found in the Fuzzy view, and a Fuzzy Logic Engine implemented in the fuzzy.spin object. ViewPort's fuzzy logic implementation consists of fuzzy maps, fuzzy rules, and fuzzy logic functions.*

Figure 11.32 is a screenshot of the ViewPort™ program running a Lunar-Lander simulation that uses an FLC. In the lower pane, you should also be able to see the three-input-variable

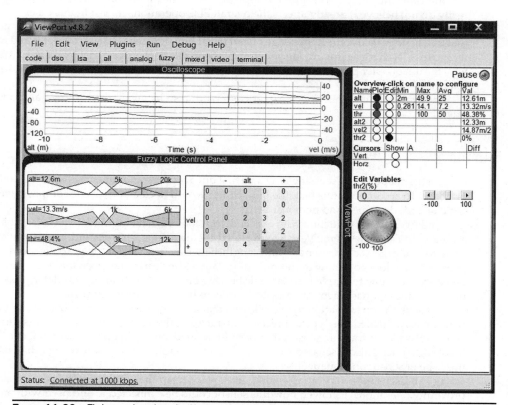

Figure 11.32   FL Lunar Lander simulation running on the ViewPort application.

	-	alt		+	
-	0	0	0	0	0
	0	0	0	0	0
vel	0	0	2	3	2
	0	0	3	4	2
+	0	0	4	4	2

Figure 11.33   Resultant rules matrix.

membership function graphs, which represent altitude, velocity, and thrust. I also extracted a portion of Figure 11.32 to show you a close-up view of the resultant rules matrix that shows all the possible outcomes for the velocity and altitude variables, as applied to the relevant rules.

Figure 11.33 shows this resultant rules matrix with the altitude and velocity both in the positive region. When you look at the shaded matrix area, it appears that the combined output would be approximately 3. It also appears that the ViewPort™ FL output does not use the centroid, or center-of-gravity approach but relies more on a weighted average. I really do not believe it greatly affects the overall FLC operation, at least in this case.

This last section concludes my AI and FL discussion and also finishes this book. I hope that you have gained some knowledge of and insight on how to build a quadcopter and how it functions. I have tried to provide you with a reasonable background that explains how and why the quadcopter performs as it does and also how to modify it to suit your own personal interests.

Remember, have fun flying the quadcopter, but also be aware of your own and others safety!

Summary

I began the chapter by discussing how a virtual quadcopter could return to its start position, or home. To accomplish this feat, the quadcopter would need an electronic compass sensor, which I both described and demonstrated.

I followed the introduction with a brief discussion of what longitudinal path lengths are and how to calculate them. Next, we explored the haversine formula and how it could be used to compute the great-circle path length between two sets of geographic coordinate pairs. We then looked at how to compute the relative true bearing between these coordinates. I next showed how to derive the magnetic bearing once the true bearing was determined, as the electronic compass mentioned earlier works only with magnetic bearings.

A discussion on formation flying followed that included specific close-proximity sensors, which partly enabled this type of precision quadcopter flying. A demonstration was shown, using a Parallax ultrasonic sensor, that permitted quadcopters to operate within 2 cm (0.79 in) of each other. I also discussed some concerns that you should be aware of regarding using this type of sensor on a quadcopter.

I next discussed a very small analog ultrasonic sensor that could also be used for close-proximity operations. I showed how this sensor could be connected to and used with one of the Parallax Propeller analog-to-digital inputs.

The next section included an introduction to LIDAR, which is a combination word for light and radar. It is a very powerful sensor system that is capable of both obstacle detection and carrying out long-range mapping operations to over 3 km (1.86 mi). LIDAR has also been used in many autonomous robotic projects.

I next discussed *fuzzy logic* (FL), which is a branch of the *artificial intelligence* (AI) field of study that is particularly well suited for quadcopter control applications. I tried to provide a somewhat comprehensive introduction to FL and *fuzzy logic control* (FLC) by using a room heating and cooling example. In this example, you saw input and output membership functions as well as the rule set. The rule set captures and encapsulates human expertise such that it can be applied to provide "intelligent" decisions based upon a given set of input values from the membership functions.

An FLC application was next shown to illustrate how FL could be applied to an operating quadcopter. I also used a Matlab® project to further illustrate quadcopter FLC operations.

The chapter concluded with an introduction to ViewPort™, which is a complementary Propeller-chip development environment that happens to have built-in FL functions. Using ViewPort™ makes Propeller FLC development very straightforward, especially because it already does most of the complex programming for you.

Index